Ultimate Exam Practice

GCSE exam secrets

Fiona C Mapp

CONTENTS

THIS BOOK AND YOUR GCSE EXAMS

Introduction

This book is designed to help you get better results.

- Look at the grade D and B candidates' answers and see if you could have done better.
- Try the exam practice questions and then look at the answers.
- Make sure you understand why the answers given are correct.
- When you feel ready, try the GCSE mock exam papers.

If you perform well on the questions in this book you should do well in the examination. Remember that success in examinations is about hard work, not luck.

What examiners look for

The obvious:

- Work which is legible, set out clearly, and easy to follow and understand; that you have used pen, not pencil, except in drawings, and that you have used the appropriate equipment.
- That drawings and graphs are neat, and graphs are labelled.
- That you always indicate how you obtain your answers.
- The right answer!

Exam technique

- Start with question 1 and work through the paper in order.
- If you cannot answer a question, leave it out and return to it at the end.
- For explanation questions, use the mark allocation to guide you on how many points to make.
- Make sure you answer the question that has been asked.
- Set your work out neatly and logically. Untidy work is not only more difficult for the examiner to read but you could also misread your own figures.
- Show all the necessary working so that you can earn the method marks even if you make a numerical slip. You need to convince the examiner that you know what you are doing.
- Do not be sloppy with algebraic notation and manipulation, brackets and negatives in particular.
- Do rough calculations to check your answers and make sure they are reasonable.
- Do not plan to have time left over at the end. If you do, use it usefully. Check you have answered all the questions, check arithmetic and read longer answers to make sure you have not made silly mistakes or missed things out.

See also Top Tips for Exam Success on page 7.

DIFFERENT TYPES OF EXAM QUESTIONS

Understand the question

It is important that you understand what the examiner means when using words like 'State', 'Find' and 'Deduce'. Here is a brief glossary to help you:

Write down, state – no working out or explanation needed.
Calculate, find, show, solve – some working out needed. Include enough working to make your method clear.
Deduce, hence – make use of an earlier answer to establish the result.
Prove – set out a concise logical argument, making the reasons clear.
Sketch – show the general shape of a graph, its relationship with the axes and points of special significance.
Draw – if a graph, plot accurately using graph paper; use geometrical instruments carefully for other diagrams.
Find the exact value – leave in fractions, surds or π. Rounded results from a calculator will not earn the marks.

Multistep questions

Multistep questions require you to obtain the answer to a problem where more than one step is required. Questions which are not multistep are sometimes called **structured.** In these, there may still be more than one step, but each stage is set out in the question. Here is an example.

Structured questions

A door wedge is in the shape of a prism with cross section VWXY.
VW = 7 cm, VY = 0.15 m, WX = 9 cm.
The width of the door wedge is 0.08 m.

(a) Convert 0.15 metres into centimetres. [1]
(b) Calculate the area of the shape VWXY. [3]
(c) Calculate the volume of the door wedge, clearly stating your units. [3]

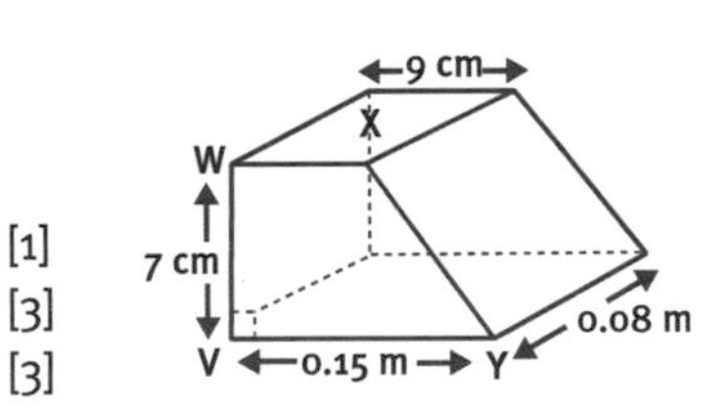

Multistep question

Calculate the volume of the door wedge, clearly stating your units. [7]

In the second form of the question it is necessary to convert all the units to either cm or m, then work out the area of cross-section and then finally the volume. The work is exactly the same, but in the second case you are not told what the intermediate steps are. You need to work out the strategy for yourself.

Using a calculator

Half of GCSE Mathematics papers must be answered without the use of calculating aids. This means that either the first paper is non-calculator, or the first section of each paper is non-calculator.
In this book, questions which may be answered with a calculator are marked with this symbol.
For the mock examination at the end of the book, the first paper is non-calculator.

WHAT IS NEEDED TO BE AWARDED A GRADE C IN MATHEMATICS

The following statements are a list of the minimum requirements of those aspects of mathematics that you are expected to know for the award of a grade C on the written examination papers. They are based on the generic descriptors that are used by all examination boards and do not include any requirements that would be covered by coursework.

A grade C candidate should be able to:

- Give reasons, justify and explain solutions to problems.
- Round to one significant figure and multiply and divide mentally.
- Solve numerical problems involving multiplication and division using a calculator efficiently.
- Understand the effects of multiplying numbers between 0 and 1.
- Understand and use the equivalences between fractions, decimals and percentages.
- Calculate using ratios and proportional changes.
- Find the nth term of a sequence.
- Multiply two expressions of the form $(x + n)$ and simplify the resulting quadratic.
- Solve equations by using trial and improvement.
- Represent inequalities using a number line.
- Form and solve linear equations.
- Manipulate simple algebraic formulae, equations and expressions.
- Solve simultaneous linear equations using an algebraic or graphical method.
- Solve problems using the angle and symmetry properties of intersecting and parallel lines.
- Understand and use Pythagoras' Theorem in two dimensions.
- Find areas and circumferences of circles.
- Calculate lengths, areas and volumes of shapes and solids.
- Enlarge shapes by a positive or fractional scale factor.
- Understand and use compound measures, such as speed.
- Construct and interpret frequency diagrams.
- Specify hypotheses and test them.
- State the modal class and estimate the mean, median and range of a set of grouped data.
- Use frequency distributions and measures of average and range to compare distributions.
- Draw a line of best fit on a scatter diagram.
- Use relative frequency as an estimate of probability.

What do I need for a grade B?

Unlike grade C, there are no criteria for the B grade. You need to get nearly all of the questions right. You should have no weaknesses in any area of the specification. The boundary mark is determined to be the same number of marks above grade C as grade D is below grade C.

HOW TO BOOST YOUR GRADE

Common areas of difficulty

These are the parts of the GCSE Intermediate tier which candidates find difficult. You will find help in revising these in Letts GCSE Success guide. The page references are given in brackets.

- Algebra, especially solving equations, inequalities, changing the subject of the formula and graph work (pages 28–34, 40–42).
- Reverse percentages, e.g. finding the cost before VAT was added (page 14).
- Ratio and proportion (page 20).
- Standard Form (page 24).
- Solving 2-D problems using Pythagoras' Theorem and Trigonometry (pages 60–66).
- Similarity (page 58).
- Loci (page 59).
- Averages of grouped data and interpreting cumulative frequency graphs (pages 82–86).

Top tips for exam success

- Practise all aspects of manipulative algebra, solving equations, rearranging formulae, expanding brackets, factorising, simplifying.
- Practise answering questions without the use of a calculator.
- Practise answering questions with more than one step to the answer (multistep questions), e.g. finding the area of a compound shape.
- Make your drawings and graphs neat and accurate.
- Practise answering questions that ask for an explanation or proof. Your answers should be concise and use mathematical terms where appropriate.
- Don't forget to check your answers, especially to see that they are reasonable. The mean weight of a group of women will not be 150 kilograms!
- Lay out your working carefully and concisely. Write down the calculations that you are going to make. You usually get marks for showing a correct method.
- Know what is on and what is not on the formula sheet before the examination.
- Make sure you can use your calculator efficiently. Write down the figures on your calculator and then make suitable rounding. Don't round the numbers during the calculation. This will often result in an inaccurate answer.
- Make sure you have read the question carefully so you give the answer the examiner wants!

CHAPTER 1

Number

To revise this topic more thoroughly, see pages 4–25 in *Letts GCSE Success Guide.*

Try this sample GCSE question and then compare your answers with the Grade D and Grade B model answers on the next page.

1 a Jacob bought a car for £12 450. He sold the car six months later for £10 200. What is his percentage loss?

....

....

.... **[3]**

b In a sale all goods are reduced by 15%.
Rebecca bought a washing machine in the sale for £314.50.
Work out the original price of the washing machine.

....

....

.... **[3]**

c Raj put £4000 in a buildings society savings account. Compound interest at 3.6% was added at the end of each year.
Calculate the total amount of money in Raj's savings account at the end of the 3 years. Give your answer to the nearest penny.

....

....

....

....

.... **[4]**

d Amy also put a sum of money into a building society savings account.
Compound interest at 4% was added at the end of each year.
Work out the number by which Amy has to multiply her sum of money to find the total amount she will have after 3 years

....

....

....

....

.... **[2]**

Total 12 marks

These two answers are at Grades D and B. Compare which one your answer is closest to and think how you could have improved it. See GCSE Success Guide page 14 for further help.

GRADE D ANSWER

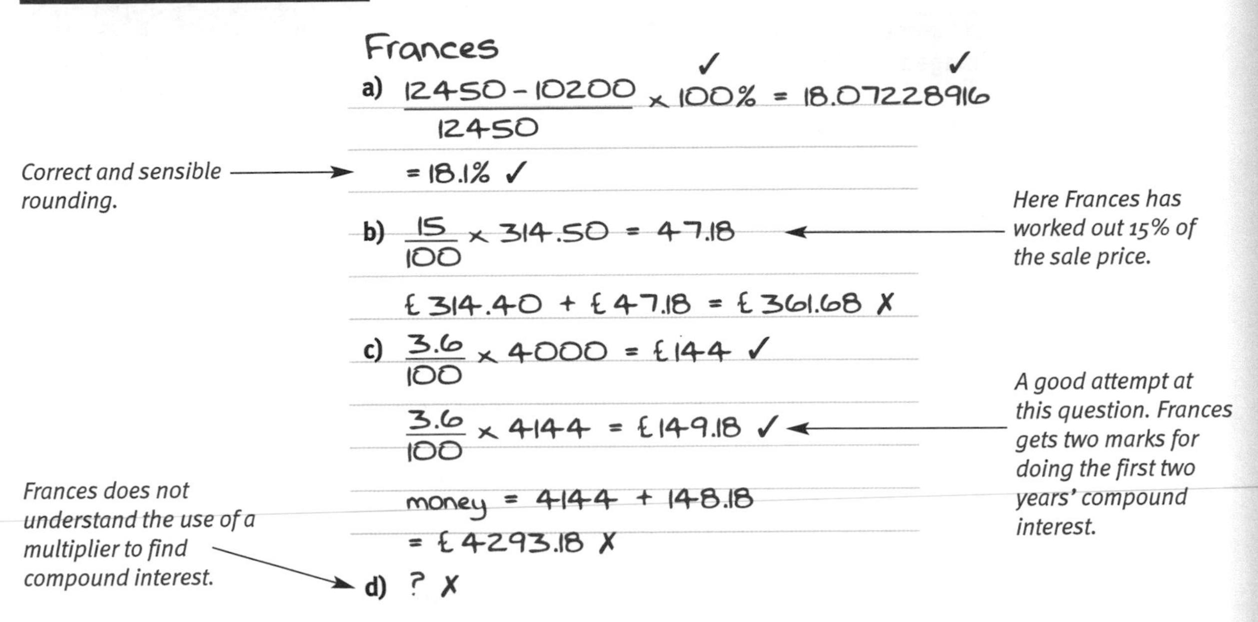

Correct and sensible rounding.

Here Frances has worked out 15% of the sale price.

A good attempt at this question. Frances gets two marks for doing the first two years' compound interest.

Frances does not understand the use of a multiplier to find compound interest.

5 marks = Grade D answer

Grade booster ····> move a D to a C

Frances can do problems involving simple percentages, but to be successful at grade C, she needs to be more competent with using multipliers. The use of multipliers for parts b), c) and d) would have made these questions more straightforward. This could lead to a grade B.

GRADE B ANSWER

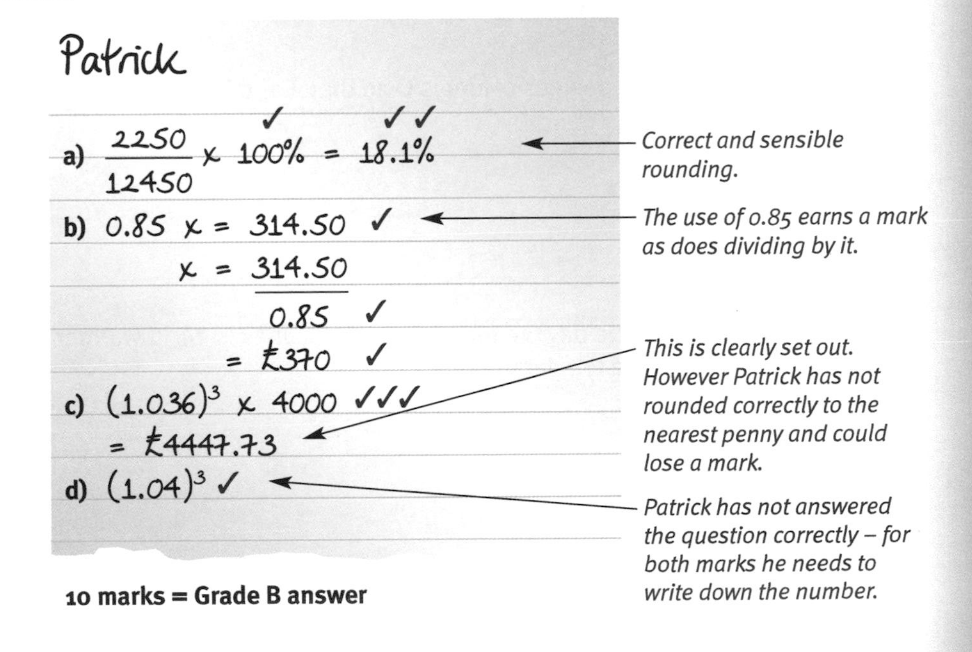

Correct and sensible rounding.

The use of 0.85 earns a mark as does dividing by it.

This is clearly set out. However Patrick has not rounded correctly to the nearest penny and could lose a mark.

Patrick has not answered the question correctly – for both marks he needs to write down the number.

10 marks = Grade B answer

Patrick has shown good competence when using percentages. He has lost marks through careless errors and not answering the questions properly.

QUESTION BANK

1 John goes on holiday to America.
The exchange rate is £1 = 1.68 US dollars.
He changed £500 into US dollars.

a) How many dollars should John get?

....

....

....dollars ②

John comes back home.
He changes 320 US dollars back into pounds. The exchange rate is the same.

b) How much money should he get? Give your answer to the nearest penny.

....

....

....£ .... ②

TOTAL 4

2 The temperature at midnight one night in December in various cities is shown in the table below.

CITY	CAIRO	LONDON	NEW YORK	OSLO
TEMPERATURE (°C)	8	−2	−7	−14

a) How many degrees colder is Oslo than Cairo?

....

....

.... ①

b) i) On the same day the temperature in Sydney is 24 °C warmer than New York.
What is the temperature in Sydney?

....

....

....

ii) How many degrees colder is London than Sydney?

....

.... ③

TOTAL 4

ANSWERS ON PAGE 21 ANSWERS ON PAGE 21 ANSWERS ON PAGE 21 ANSWERS ON PAGE 21

3 Work out 30% of £2600.

(2)

TOTAL 2

4 Arrange these numbers in order of size, smallest first.

30%, $\frac{2}{5}$, 0.7, $\frac{7}{8}$, 65%, $\frac{1}{3}$

(3)

TOTAL 3

5 Rupinder buys 25 calculators at £6.80 each.

a) Work out the total cost of the 25 calculators.

(3)

6 out of the 25 calculators do not work.

b) Write 6 out of 25 as a percentage.

(2)

TOTAL 5

6 Work out

a) $3\frac{1}{4} + 1\frac{2}{5}$

(3)

b) $1\frac{7}{8} \times 2\frac{1}{2}$

(3)

TOTAL 6

ANSWERS ON PAGE 21 ANSWERS ON PAGE 21 ANSWERS ON PAGE 21 ANSWERS ON PAGE 21

7 Work out

a) $5\frac{1}{4} - 3\frac{1}{5}$

....

.... ③

b) $\frac{4}{7} \div 1\frac{1}{4}$

....

.... ③

TOTAL 6

8 The diagram shows the distances in miles between some villages.
The villages are labelled with letters.

$4\frac{1}{4}$ $3\frac{1}{2}$

A B C D E

The distance from village A to village B is $4\frac{1}{4}$ miles.

The distance from village B to village C is $3\frac{1}{2}$ miles.

a) Work out the total distance in miles between village A and village C.

....

.... ②

b) The distance from village A to village D is $10\frac{1}{2}$ miles. Work out the distance in miles, from village C to village D.

....

....

.... ②

c) The distance from village D to village E is three-quarters of the distance from village B to village C. Work out the distance in miles from village D to village E.

....

....

.... ③

TOTAL 7

ANSWERS ON PAGE 21 ANSWERS ON PAGE 21 ANSWERS ON PAGE 21 ANSWERS ON PAGE 21

9 Charlotte wins £16 000 on the lottery. She keeps a quarter of the money and gives a fifth to charity. She then divides the remaining money between her four friends. How much does each friend receive?

....

....

....

.... (4)

TOTAL 4

---

10 Jonathan wants to buy a covet dvd player
Covet dvd players are sold in four different shops

Robbies	Electricals	Dillmons
$\frac{1}{3}$ off	25% off	£215
usual price of £340	usual price of £310	plus VAT at $17\frac{1}{2}$%

a) Find the difference between the maximum and minimum prices Jonathan could pay for a Covet dvd player.

....

....

....

....

....

.... (7)

b)

Electrical World £267.90 including VAT

The price of the Covet dvd player in Electrical World is £267.90.
This includes VAT at $17\frac{1}{2}$%.

Work out the cost of the Covet dvd player before VAT is added.

....

....

.... (3)

TOTAL 10

ANSWERS ON PAGE 21 ANSWERS ON PAGE 21 ANSWERS ON PAGE 21 ANSWERS ON PAGE 21

11 In a sale, all the normal prices are reduced by 12%.
The normal price of a coat is £125.
Bola buys the coat in the sale.

a) Work out the sale price of the coat.

....

....

.... ②

In the same sales, Nigel pays £19.80 for a shirt.

b) Calculate the normal price of the shirt.

....

....

....

.... ③

TOTAL 5

---

12 £6000 is invested for 3 years at 5% per annum compound interest.
Work out the total interest earned over the three years.

....

....

....

.... ③

TOTAL 3

---

13 Poppy bought a computer for £1250.
Each year the computer depreciates by15%.

Work out its value two years after she bought it.

....

....

.... ③

TOTAL 3

---

14 Jerry buys a flat for £82 000 in 2000. During 2001 house prices rise by 8% and then by a further 13% in 2002.
How much is Jerry's flat worth at the end of 2002?

....

.... ③

TOTAL 3

ANSWERS ON PAGE 21 ANSWERS ON PAGE 21 ANSWERS ON PAGE 21 ANSWERS ON PAGE 21

15. Emily buys a car for £12 500. Two years later she sells the car for £7250.
Work out the percentage loss.

....

....

.... ③

TOTAL 3

---

16. Sam bought 6 identical pens for £4.68.
Work out the cost of 11 of these pens.

....

....

....

.... ②

TOTAL 2

---

17. Colin gave his three children a total of £72.05.
The money was shared in the ratio 5 : 4 : 2.
Daisy had the largest share.
Work out how much Colin gave to Daisy.

....

....

.... ③

TOTAL 3

---

18. It takes 3 builders 5 days to build a garage. The builders all work at the same rate.
How long would it take 10 builders to build the garage.

....

....

.... ③

TOTAL 3

ANSWERS ON PAGE 21 ANSWERS ON PAGE 21 ANSWERS ON PAGE 21 ANSWERS ON PAGE 21

19 The ingredients of 8 small cakes are:

300 grams of self raising flour
150 grams of butter
250 grams of sugar
2 eggs.

Imran is making 20 small cakes. Write down the amounts of ingredients he will need.

.... grams of self raising flour

.... grams of butter

.... grams of sugar

.... eggs

(4)

TOTAL 4

---

20 Toothpaste is sold in three different sized tubes.

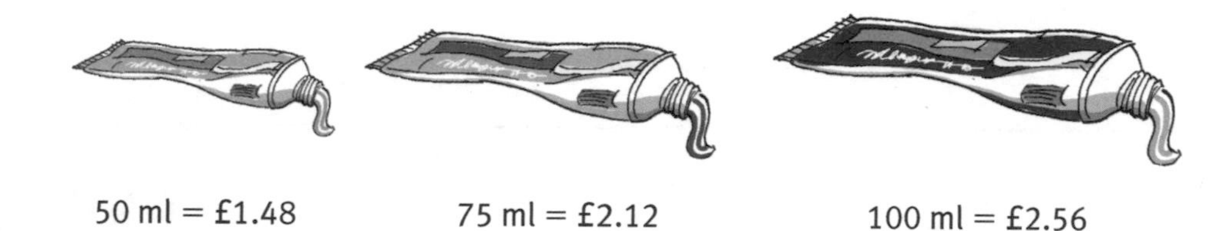

50 ml = £1.48   75 ml = £2.12   100 ml = £2.56

Which of the tubes of toothpaste is the best value for money?
You must show full working out in order to justify your answer.

....

....

....

....

....

....

....

(4)

TOTAL 4

21 Work out an estimate for

$$\frac{59.43201}{0.20141}$$

....

....

....

.... ②

TOTAL 2

22 Work out an estimate for the value of

$$\frac{49.36 - 3 \times 10.06}{}$$

Give your answer as a fraction in its simplest form.

....

....

.... ③

TOTAL 3

23 Work out the value of

$$\sqrt{\frac{64.5 \times 279}{729 + 468}}$$

Give your answer correct to 3 significant figures.

....

....

.... ③

TOTAL 3

24 Calculate

$$\frac{2.79 \times 4.63^2 - 2.71 \sin 30^\circ}{\sqrt{8.29} + 4.627}$$

....

....

....

.... ③

TOTAL 3

ANSWERS ON PAGE 21 ANSWERS ON PAGE 21 ANSWERS ON PAGE 21 ANSWERS ON PAGE 21

25 a) Write 28 and 42 as products of their prime factors.

....

....

....

.... (4)

b) i) Find the highest common factor of 28 and 42.

....

.... (2)

ii) Find the lowest common multiple of 28 and 42.

....

....

.... (2)

TOTAL 8

---

26 The number 360 can be written as $2^a \times 3^b \times 5^c$.
Calculate the values of a, b and c.

....

....

.... (3)

TOTAL 3

---

27 a) Evaluate the following.

i) 6^0

....

ii) 4^{-2}

....

iii) $25^{1/2}$

.... (3)

b) Write the following as a single power of 4.

$$\frac{4^5 \times 4^3}{(4^2)^3}$$

....

.... (2)

TOTAL 5

ANSWERS ON PAGE 21 ANSWERS ON PAGE 21 ANSWERS ON PAGE 21 ANSWERS ON PAGE 21

28 Simplify

a) i) $36^{\frac{1}{2}}$ ii) 7^{-2} iii) 16^{0}

.... ③

b) Six multiplied by the reciprocal of thirty-six.

.... ①

TOTAL 4

---

29 $2.64 \times 350 = 924$.
Use this result to write down the answer to

i) 2.64×35

....

ii) 2.64×3.5

....

iii) 0.264×0.35

.... ③

TOTAL 3

---

30 Write as simply as possible, giving your answer in standard form.

a) $(4 \times 10^{6}) \times (2 \times 10^{-3})$

.... ②

b) $(6 \times 10^{-5}) \times (7 \times 10^{8})$

.... ②

c) $\dfrac{3.6 \times 10^{8}}{7.2 \times 10^{-3}}$

.... ②

TOTAL 6

---

31 Write these numbers in standard form.

i) 2 700 000 000 ii) 0.00000729

.... ②

ii) A company is worth 7×10^{12} pounds.
Write this as an ordinary number.

.... ①

TOTAL 3

Number

32 a) Work out the value of $\frac{5 \times 10^{-7}}{2 \times 10^{-4}}$

Give your answer in standard form.

....

.... ②

b) The distance from the Earth to one of the stars is 3.72 light years.
1 light year = 9.461×10^{12} km.

Work out the distance in kilometres from Earth to the star.
Give your answer in standard form, correct to 3 significant figures.

....

....

.... ②

TOTAL 4

33 32 000 000 onion seeds weigh 10 grams.
Each onion seed weighs the same.

a) Write the number 32 000 000 in standard form.

....

.... ①

b) Calculate the weight in grams of one onion seed.
Give your answer in standard form, correct to 2 significant figures.

....

....

.... ②

TOTAL 3

ANSWERS ON PAGE 21 ANSWERS ON PAGE 21 ANSWERS ON PAGE 21 ANSWERS ON PAGE 21

QUESTION BANK ANSWERS

1 a) 840 US dollars ❷
b) £190.48 ❷

EXAMINER'S TIP

When changing the dollars back to pounds divide by the exchange rate.

2 a) 22°C ❶
b) i) 17°C ❶
ii) 19 degrees ❷

EXAMINER'S TIP

For part (a) remember subtracting a negative is the same as adding. $8 - (-14) = 8 + 14 = 22$.

3 10% = £260
30% = 3 × £260
= £780 ❷

EXAMINER'S TIP

On the non-calculator paper working out 10% first makes the calculation easier.

4 30%, $\frac{1}{3}$, $\frac{2}{5}$, 65%, 0.7, $\frac{7}{8}$ ❸

EXAMINER'S TIP

Check that you have included all the values.

5 a) £170 ❸
b) $\frac{6}{25} = \frac{24}{100} = 24\%$ ❷

EXAMINER'S TIP

For part (b) change $\frac{6}{25}$ into an equivalent fraction with a denominator of 100.

6 a) $4\frac{13}{20}$ ❸
b) $4\frac{11}{16}$ ❸

EXAMINER'S TIP

It is best to write your answer as a mixed number.

7 a) $2\frac{1}{20}$ ❸
b) $\frac{16}{35}$ ❸

EXAMINER'S TIP

When dividing (or multiplying) change the mixed numbers into top heavy fractions.

8 a) $7\frac{3}{4}$ miles ❷
b) $2\frac{3}{4}$ miles ❷
c) $2\frac{5}{8}$ miles ❸

EXAMINER'S TIP

Treat these questions as addition/subtraction of fractions.

9 $\frac{1}{4}$ of £16000 = £4000

$\frac{1}{5}$ of £16000 = £3200

money left = £8800
Each friend receives £2200 ❹

EXAMINER'S TIP

In this multistep question be systematic and show clearly your working out.

10 Robbie's $\frac{1}{3} \times 340$ = £113.33
price=
= £226.67

Electricals 25% × 310 = £77.50
price = £310 − 77.50
= £232.50

Dillmons 17.5% of £215 = £37.63
price = 215 + 37.63
= £252.63

Difference in prices £252.63 − £226.67
= £25.96 ❼

b) $\frac{£267.90}{1.175}$ = original price

original price = £228 ❸

EXAMINER'S TIP

In part (a) remember to answer the question, i.e. the difference between the maximum and minimum price.

11 a) £110 (2)

b) $\frac{19.80}{0.88} = £22.50$ (3)

EXAMINER'S TIP

Remember to divide by 0.88.

12 $6000 \times (1.05)^3$
$= £6945.75$
Interest = £945.75 (3)

EXAMINER'S TIP

Check that you have answered the question – it is the interest that is required.

13 £903.13 (3)

EXAMINER'S TIP

A quick way of working out the answer is to multiply the original price by 0.85^2

14 $82000 \times 1.08 \times 1.13$
£100072.80
£100073 (3)

EXAMINER'S TIP

This time the multiplier changes each year so the original price is multiplied by 1.08 and 1.13.

15 42% (3)

EXAMINER'S TIP

When calculating a percentage change, remember to divide by the original amount.

16 £8.58 (2)

EXAMINER'S TIP

With questions like these it is easier to work out the cost of one item first.

17 £32.75 (3)

EXAMINER'S TIP

Work out what one part is worth, which then makes working out what each child receives easy.

18 3 builders take 5 days.
1 builder takes $5 \times 3 = 15$ days.
10 builders take $15 \div 10 = 1\frac{1}{2}$ days. (3)

EXAMINER'S TIP

Remember that one builder will take longer than three builders to do the job so we multiply and not divide in this case.

19 750 grams of self raising flour
375 grams of butter
625 grams of sugar
5 eggs (4)

EXAMINER'S TIP

The multiplier is 2.5, so multiply all ingredients by 2.5.

20 50 ml = £1.48 so 1 ml = 2.96p
75 ml = £2.12 so 1 ml = 2.83p
100 ml = £2.56 so 1 ml = 2.56p (4)
The 100 ml of toothpaste is the best value for money.

EXAMINER'S TIP

You must show full and clear working out in order to get full marks.

21 $\frac{60}{0.2} = 300$ (2)

EXAMINER'S TIP

When estimating remember to round the numbers to 1 significant figure

22 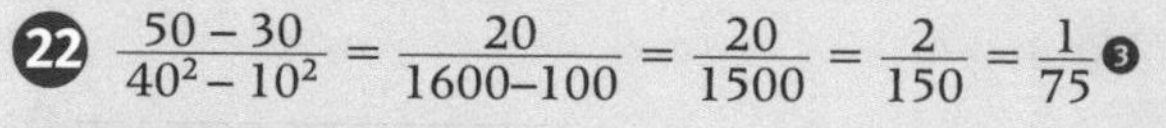
$\frac{50-30}{40^2-10^2} = \frac{20}{1600-100} = \frac{20}{1500} = \frac{2}{150} = \frac{1}{75}$ (3)

EXAMINER'S TIP

You will lose a mark if you do not simplify the fraction fully.

23 3.88 (3)

EXAMINER'S TIP

Remember to round your answer to 3 significant figures as instructed.

24 $\frac{58.453951}{3.5940228} = 16.26$ (2dp) (3)

EXAMINER'S TIP

Show intermediate working in order to gain some method marks.

FOR MORE INFORMATION ON THIS TOPIC ... SEE GCSE SUCCESS GUIDE ... PAGES 4–25

25 a) $28 = 2 \times 2 \times 7$ ❹

$42 = 2 \times 3 \times 7$

b) i) HCF $2 \times \boxed{2} \times \boxed{7}$

$\boxed{2} \times \boxed{7} \times 3$ ❷

HCF $= 2 \times 7$

$= 14$

ii) LCM $= 2 \times 2 \times 7 \times 3$

$= 84$ ❷

EXAMINER'S TIP

When dividing to find factors, be systematic e.g. start with 2. When no more factors of 2, try 3 and so on.

26 $360 = 2^3 \times 3^2 \times 5^1$

$a = 3$

$b = 2$

$c = 1$ ❸

EXAMINER'S TIP

Divide to find the factors as before, then read the indices.

27 a) i) 1 ❶

ii) $\frac{1}{4^2} = \frac{1}{16}$ ❶

iii) $\sqrt{25} = \pm 5$ ❶

b) $\frac{4^5 \times 4^3}{(4^2)^3} = \frac{4^8}{4^6} = 4^2$ ❷

EXAMINER'S TIP

A fractional index means a root.

28 a) i) ± 6 ii) $^1/_{49}$ iii) 1 ❸

b) $^1/_6$ ❶

EXAMINER'S TIP

When the index is negative you take the reciprocal, for example $7^{-2} = \frac{1}{7^2}$

29 a) i)92.4 ❶

ii) 9.24 ❶

iii) 0.0924 ❶

EXAMINER'S TIP

Think carefully about the place values of the numbers.

30 a) 8×10^3 ❷

b) 4.2×10^4 ❷

c) 5×10^{10} ❷

EXAMINER'S TIP

Make sure that your answers are written in standard form.

31 i) 2.7×10^9 ❷

ii) 7.29×10^{-6}

iii) £7 000 000 000 000 ❶

EXAMINER'S TIP

For small numbers the power of 10 is negative.

32 a) 2.5×10^{-3} ❷

b) 3.52×10^{13} km. ❷

EXAMINER'S TIP

Make sure that you know how to input standard form on your calculator.

33 a) 3.2×10^7 ❶

b) 3.125×10^{-7} grams

$= 3.1 \times 10^{-7}$ grams (2 sf) ❷

EXAMINER'S TIP

Check that you have rounded the answer in part (b) to 2 significant figures.

CHAPTER 2

Algebra

To revise this topic more thoroughly, see pages 28–43 in *Letts GCSE Success Guide.*

Try this sample GCSE question and then compare your answers with the Grade D and Grade B model answers on the next page.

1 a (i) Simplify $a + a + 2a$

.... **[1]**

(ii) Simplify $3a + 4b - 4a + 2b$

.... **[1]**

(iii) Expand $5a(2a + 3)$

.... **[1]**

b Expand and simplify

$3(x - 2) - 2(x - 2)$

.... **[2]**

c i) Simplify $a^3 \times a^4$

.... **[1]**

ii) Simplify $\frac{6a^2b^3}{2a^3b}$

.... **[1]**

d Expand and simplify

$(x + 4)(2x - 3)$

.... **[2]**

e (i) Factorise $8a - 4$

.... **[1]**

(ii) Factorise completely

$6p^3 - 9p^2q$

.... **[2]**

f (i) Factorise $x^2 + 3x - 10$

.... **[2]**

(ii) Hence or otherwise

solve $x^2 + 3x - 10 = 0$

.... **[2]**

Total 16 marks

These two answers are at Grades D and B. Compare which one your answer is closest to and think how you could have improved it. See GCSE Success Guide pages 30–31 for further help.

GRADE D ANSWER

Sanjay

a (i) $4a$ ✓

(ii) $7a + 6b$ ✗

(ii) $10a^2 + 15a$ ✓

b $3x - 6 - 2x + 4$ ✓

$= x - 10$ ✗

c (i) a^7 ✓

(ii) $\frac{3b^2}{a}$ ✓

d $(x + 4)(2x - 3)$

$= 2x^2 - 3x + 8x - 12$ ✓

$= 2x^2 + 11x - 12$ ✗

e (i) $4(2a - 1)$ ✓

(ii) $3p(2p^2 - 3pq)$ ✗

f (i) $x^2 + 3x - 10$

$= x(x + 3) - 10$ ✗

(ii) $x = ?$ ✗

Correctly remembered to multiply both terms with 5a.

Sanjay has ignored the negative 4a and treated it as 4a.

One mark is obtained for correctly multiplying out the brackets.

Sanjay has a good grasp of indices.

Sanjay has problems with collecting like terms, when there is a negative term.

Sanjay obviously has difficulty with factorising and solving a quadratic equation. This is grade B work.

7 marks = Grade D answer

Grade booster ⟶ move a D to a C

Sanjay has quite a good grasp of basic algebra but he has lost several marks because he ignores the negative sign when collecting like terms. He needs to practise factorisation, which is a more difficult topic at Intermediate level.

GRADE B ANSWER

Charlotte

a (i) $4a$ ✓

(ii) $6b - a$ ✓

(iii) $10a^2 + 15a$ ✓

b $3(x - 2) - 2(x - 2)$

$= 3x - 6 - 2x + 4$ ✓ $= x - 2$ ✓

c (i) a^7 ✓

(ii) $\frac{3b^2}{a}$ ✓

d $(x + 4)(2x - 3)$

$= 2x^2 - 3x + 8x - 12$ ✓

e (i) $4(2a - 1)$ ✓

(ii) $3p(2p^2 - 3pq)$ ✗

f (i) $x^2 + 3x - 10$

$(x - 2)(x + 5)$ ✓✓

Check : $x^2 + 5x - 2x - 10$

(ii) $(x - 2) = 0 \quad x = +2$ ✓

$(x + 5) = 0 \quad x = 5$ ✗

A grade B candidate would be expected to get this right.

All correct and well laid out.

An answer of $3a^{-1}b^2$ would be correct.

Charlotte has lost a mark here because she has not simplified the expression.

Charlotte has not factorised completely. The correct answer is $3p^2(2p - 3q)$.

Factorising the quadratic is correct and it is nice to see that Charlotte has checked her answer.

Another careless error, Charlotte has forgotten to write -5.

12 marks = Grade B answer

Charlotte has shown good competence in algebra and should get a grade B if the rest of the paper is of a similar standard. She must make sure that she has answered the question properly.

QUESTION BANK

1 Rena earns r pounds per hour.
She works for x hours.
She also earns a bonus of y pounds.

Write down a formula for the total amount she earns, t pounds.

....

....

.... (3)

TOTAL 3

2 a) Simplify

i) $3a + 2a - a$

....

ii) $5p - 3r + 2p - 6r$

....

.... (2)

b) Find the value of

i) $2a + 3b$ When $a = 4$ and $b = 6$

....

....

ii) $3r - 6s$ When $r = 5$ and $s = 3$

....

.... (4)

TOTAL 6

ANSWERS ON PAGE 40 ANSWERS ON PAGE 40 ANSWERS ON PAGE 40 ANSWERS ON PAGE 40

3 a) Simplify fully

$$5n + 8n - 12n$$

.... ①

b) Simplify fully

$$3r - 2p + 6r - 10p$$

.... ①

c) Multiply out

$$3(2a + 4)$$

.... ①

TOTAL 3

---

4 $y = 3x - 2$

a) Find the value of x when $y = 4$.

....

.... ②

b) i) Simplify $ab + ab + ab$

.... ①

ii) Simplify $w^2 + 2w^2$

.... ①

TOTAL 4

---

5 $A = \dfrac{B\,(C + 40)}{20}$

$B = -2$
$C = 10$

Work out the value of A.

....

....

.... ③

TOTAL 3

ANSWERS ON PAGE 40 ANSWERS ON PAGE 40 ANSWERS ON PAGE 40 ANSWERS ON PAGE 40

6 $p = 2q + r$

a) Work out the value of p when $q = -4$ and $r = 2.5$.

....

....

.... ②

b) Work out the value of q when $p = 20$ and $r = -4$.

....

.... ②

TOTAL 4

7 Here are the first four terms of a simple sequence:

3, 7, 11, 15

a) Write down the next two numbers of the sequence.

.... ①

b) Write down, in terms of n, an expression for the nth term of this sequence.

....

.... ②

TOTAL 3

8 Here are the first five terms of a number sequence:

5, 8, 11, 14, 17

Write down, in terms of n, an expression for the nth term of the sequence.

....

.... ②

TOTAL 2

ANSWERS ON PAGE 40 ANSWERS ON PAGE 40 ANSWERS ON PAGE 40 ANSWERS ON PAGE 40

9 a) Simplify

i) $a^3 \times a^4$

.... ①

ii) n^3

.... ①

iii) $\dfrac{p^4 \times p^2}{p}$

.... ①

b) Simplify these expressions

i) $5ab^2 \times 2a^2b$

.... ①

ii) $12a^2b^4$

.... ②

iii) $(4a^2b)^2$

....

.... ②

TOTAL 8

---

10 Simplify these expressions

a) $2a \times 4b$

....

.... ①

b) $t^5 \times t^6$

....

.... ①

c) $\dfrac{4m^6 \times 3m^2}{2m^3}$

....

.... ①

d) $(2a^3b^5)^2$

....

.... ①

TOTAL 4

ANSWERS ON PAGE 40 ANSWERS ON PAGE 40 ANSWERS ON PAGE 40 ANSWERS ON PAGE 40

11 a) Expand

$3(x - 4)$

....

.... ①

b) Expand and simplify

$3(2x - 1) - 2(x - 2)$

....

....

.... ②

c) Expand and simplify

$(x - 2)(3x - 5)$

....

....

.... ②

d) Expand and simplify

$(x - 5)^2$

....

.... ②

e) Expand and simplify

$(2x - 3y)(3x + 2y)$

....

....

.... ②

TOTAL 9

ANSWERS ON PAGE 40 ANSWERS ON PAGE 40 ANSWERS ON PAGE 40 ANSWERS ON PAGE 40

12 a) Solve the equation

$5a - 4 = 6$

....

....

.... ②

b) Solve the equation

$15n + 3 = 5n - 7$

....

....

.... ②

c) Solve the equation

$5(n - 3) = 20$

....

....

.... ③

d) Solve the equation

$\frac{18 + 2x}{5} = 3x$

....

....

.... ③

TOTAL 10

13 a) Solve the equation

$3p - 7 = -2$

....

....

.... ②

b) Solve the equation

$2(3 - x) = 8(x + 6)$

....

....

.... ③

c) Solve the equation

$$\frac{3p - 1}{3} + \frac{p + 4}{2} = \frac{1}{6}$$

....

....

....

....

....

.... ④

TOTAL 9

14 The angles of a quadrilateral are marked on the diagram.

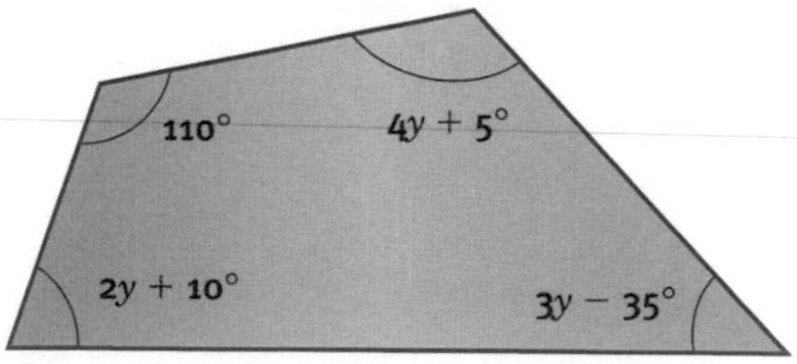

110°

$4y + 5°$

$2y + 10°$

$3y - 35°$

Form an equation in y and solve it to find the angles of the quadrilateral.

....

....

....

.... ④

TOTAL 4

ANSWERS ON PAGE 40 ANSWERS ON PAGE 40 ANSWERS ON PAGE 40 ANSWERS ON PAGE 40

15 Show that $(a - 1)^2 + a + (a - 1)$ simplifies to a^2.

....

....

....

....

.... (4)

TOTAL 4

---

16 The equation

$x^3 - 5x = 10$

has a solution between $x = 2$ and $x = 3$.
Use a method of trial and improvement to find this solution.
Give your answer correct to 1 decimal place.
You must show all your working.

....

....

....

.... (4)

TOTAL 4

---

17 n is an integer such that $-3 \leq 2n + 1 < 4$

a) List all the possible values of n.

....

....

.... (3)

b) Solve the inequality

$5 + p \geqslant 3p - 4$

....

....

.... (2)

TOTAL 5

ANSWERS ON PAGE 40 ANSWERS ON PAGE 40 ANSWERS ON PAGE 40 ANSWERS ON PAGE 40

18 Solve these inequalities.

a) $2x - 5 < 10$

....

.... ②

b) $-3 \leqslant 2x + 1 < 4$

....

....

.... ②

TOTAL 4

19 Solve the simultaneous equations

$4p + 3q = 6$
$2p - 3q = 12$

....

....

....

.... ④

TOTAL 4

20 Solve the simultaneous equations

$5c + 4d = 23$
$3c - 5d = -1$

....

....

....

.... ④

TOTAL 4

21 Factorise completely

a) $6a + 3$

....

.... ①

b) $4x^2 + 8xy$

....

.... ②

c) $(x + y)^2 - 3(x + y)$

....

.... ②

d) Factorise and simplify

$\frac{5x + 10}{x + 2}$

....

.... ②

e) Factorise

$x^2 - 36$

....

.... ②

TOTAL 9

22 a) Solve the equation

$x^2 - 5x + 6 = 0$

....

.... ②

b) Solve the equation

$x^2 - x - 12 = 0$

....

.... ②

TOTAL 4

ANSWERS ON PAGE 40 ANSWERS ON PAGE 40 ANSWERS ON PAGE 40 ANSWERS ON PAGE 40

23 $v = u + at$

a) Rearrange the formula to make t the subject.

....

.... ②

b) $E = \frac{mv^2}{2}$

Rearrange the formula to make v the subject.

....

.... ②

TOTAL 4

---

24 William went for a ride on his bike.
He rode from his home to the shops.
The travel graph shows this part of the trip.

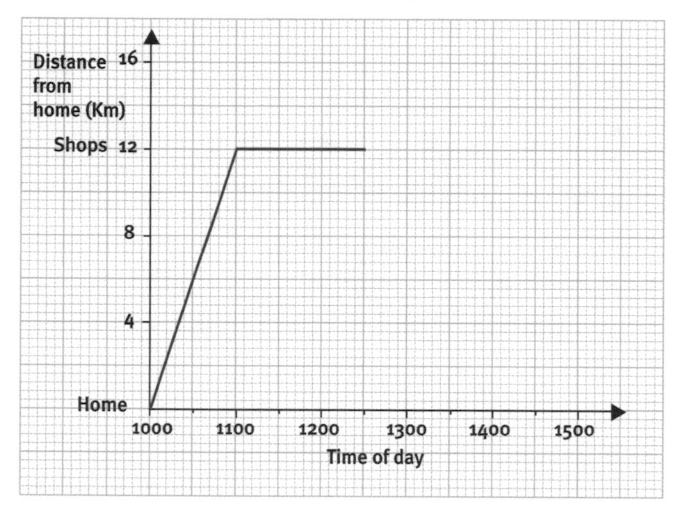

a) Find William's average speed between 1000 and 1100.

....

.... ②

b) What happened to William between 1100 and 1230?

....

....

.... ①

William started to travel back to his home at 1230.
He travelled at a speed of 8 km per hour for 45 minutes.
He then had a puncture so he stopped and repaired it. This took 15 minutes.
He then continued to travel home at 12 km per hour.

c) Complete the travel graph. ③

TOTAL 6

ANSWERS ON PAGE 40 ANSWERS ON PAGE 40 ANSWERS ON PAGE 40 ANSWERS ON PAGE 40

25 Water is being poured into these containers at a rate of 250 ml per second.
The graphs below show how the height of the water changes with time.
Match the containers with the graphs.

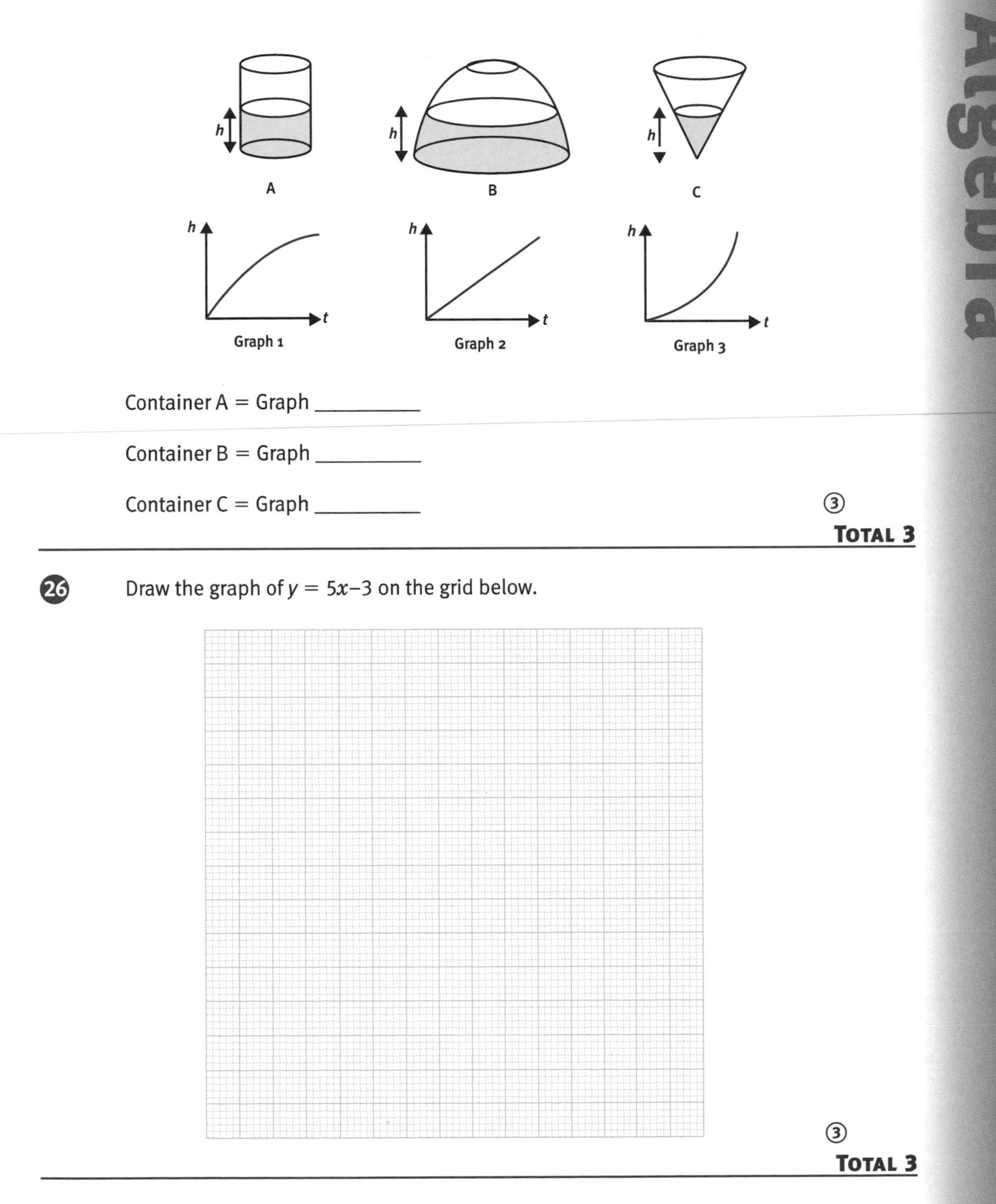

Container A = Graph __________

Container B = Graph __________

Container C = Graph __________ ③

TOTAL 3

---

26 Draw the graph of $y = 5x-3$ on the grid below.

③

TOTAL 3

27

Here are the equations of 5 straight lines.They are labelled from A to E.

A	$y = 5 - 2x$	
B	$y = 3x - 4$	
C	$2y + 4x = 7$	
D	$3x + y = 6$	
E	$6y - 2x = 5$	

Put ticks in the table to show the two lines that are parallel.

②

TOTAL 2

The diagram shows the graphs of the equations

$x + y = 5$
$y = 2x - 1$

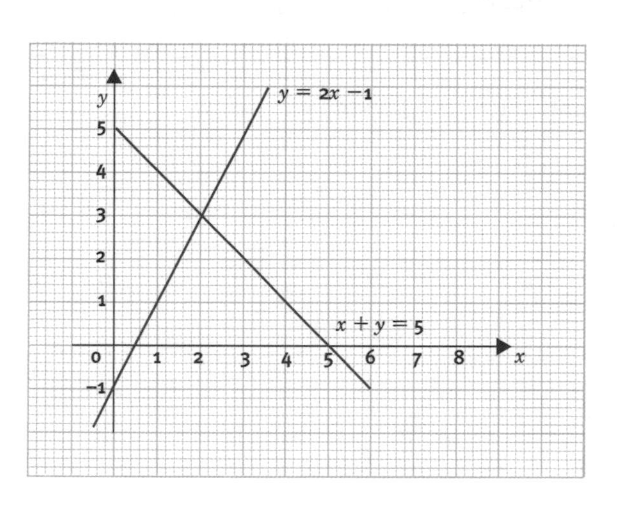

a) Use the diagram to solve the simultaneous equations

$x + y = 5$
$y = 2x - 1$

....

.... ②

b) On the grid above, shade in the region satisfied by the inequalities

$x + y \leq 5$
$y \geqslant 2x - 1$

②

TOTAL 4

29 a) Complete the table of values for the graph of $y = x^3 + 5$.

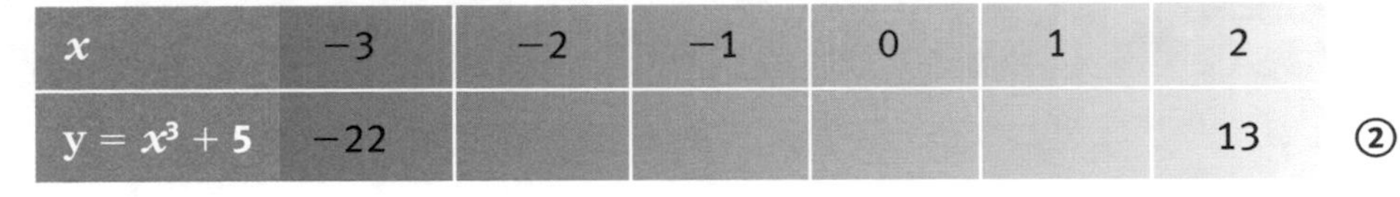

x	-3	-2	-1	0	1	2
$y = x^3 + 5$	-22					13

②

b) On the grid draw the graph of $y = x^3 + 5$. ②

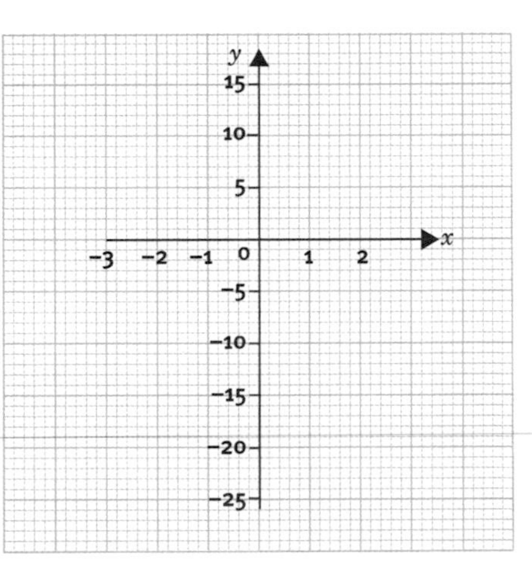

c) Use your graph to find

i) an estimate of the solution of the equation $x^3 + 5 = 0$.

....

.... ①

ii) an estimate of the solution of the equation $x^3 + 5 = -6$

....

.... ②

TOTAL 7

---

30 Match the graphs with the equations.

i) $y = x^2 - 8$ ii) $y = 1 - 2x$ iii) $xy = 4$ iv) $y = 3x - 4$

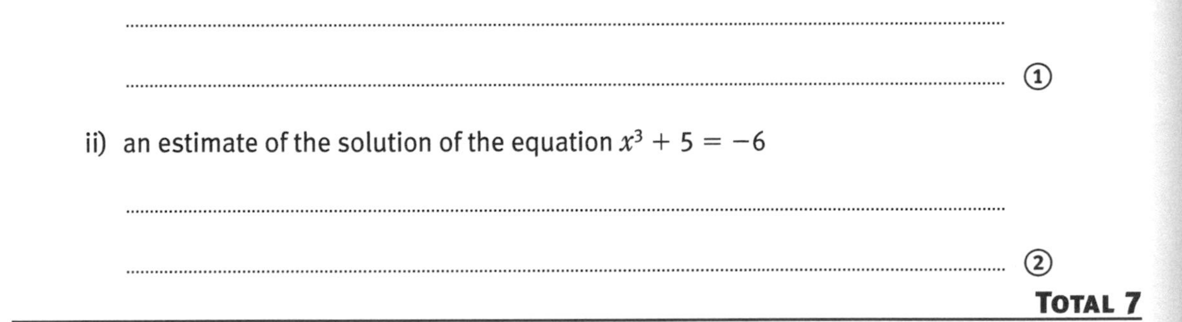

A = ... B = ... C = ... D = ... ④

TOTAL 4

ANSWERS ON PAGE 40 ANSWERS ON PAGE 40 ANSWERS ON PAGE 40 ANSWERS ON PAGE 40

QUESTION BANK ANSWERS

1 $t = rx + y$ (3)

EXAMINER'S TIP

You are asked for a formula so you must write 't='.

2 a) i) $4a$ (1)
ii) $7p - 9r$ (1)
b) i) 26 (2)
ii) -3 (2)

EXAMINER'S TIP

Beware of the negative sign in (a) (ii)

3 a) n (1)
b) $9r - 12p$ (1)
c) $6a + 12$ (1)

EXAMINER'S TIP

When multiplying out the bracket in (c), both terms must be multiplied by 3.

4 a) 2 (2)
b) i) $3ab$ (1)
ii) $3w^2$ (1)

EXAMINER'S TIP

$w^2 + 2w^2 = 3w^2$ in part (b) (ii). Do not get confused with multiplying and write $2w^4$.

5 $A = \frac{B\,(C + 40)}{20}$

$A = \frac{-2\,(10 + 40)}{20}$

$A = \frac{-2 \times 50}{20}$

$A = -5$ (3)

EXAMINER'S TIP

To earn method marks show the substitution and full working out.

6 a) -5.5 (2)
b) 12 (2)

7 a) 19, 23 (1)
b) $4n - 1$ (2)

EXAMINER'S TIP

Check that the nth term is correct by substituting a value of n back into the expression.

8 $3n + 2$ (2)

9 a) i) a^7 (1)
ii) n^{-3} or $\frac{1}{n^3}$ (1)
iii) p^5 (1)
b) i) $10a^3b^3$ (1)
ii) $\frac{4b^3}{a^3}$ (2)
iii) $16a^4b^2$ (2)

EXAMINER'S TIP

For (b) (iii) all parts of the bracket must be squared.

10 a) $8ab$ (1)
b) t^{11} (1)
c) $6m^5$ (1)
d) $4a^6b^{10}$ (1)

11 a) $3x - 12$ (1)
b) $4x + 1$ (2)
c) $3x^2 - 11x + 10$ (2)
d) $(x - 5)^2 = (x - 5)\,(x - 5)$
$= x^2 - 10x + 25$ (2)
e) $6x^2 - 5xy - 6y^2$ (2)

EXAMINER'S TIP

Remember that $(x - 5)^2$ means $(x - 5)\,(x - 5)$.

12 a) $a = 2$ (2)
b) $n = -1$ (2)
c) $n = 7$ (3)
d) $\frac{18 + 2x}{5} = 3x$

$18 + 2x = 15x$

$18 = 13x$

$x = \frac{18}{13} = 1\frac{5}{13}$ (3)

EXAMINER'S TIP

In part (d) remove the denominator by multiplying both sides by 5.

FOR MORE INFORMATION ON THIS TOPIC ... SEE GCSE SUCCESS GUIDE ... PAGES 28–43

13 a) $p = {}^5/_3 = 1^2/_3$ ❷
b) $x = -4.2$ ❸
c) $p = -1$ ❹

EXAMINER'S TIP

Remember when adding fractions they need to have a common denominator.

14 $4y + 5° + 3y - 35° + 2y + 10° + 110° = 360°$
$9y + 90° = 360°$
$9y = 360° - 90°$
$9y = 270°$
$y = 30°$
Angles in quadrilateral : 110°, 125°, 70°, 55° ❹

EXAMINER'S TIP

Form the equation first, collect like terms and then solve as normal.

15 $(a-1)^2 + a + a - 1$
$= a^2 - 2a + 1 + a + a - 1$
$= a^2 - 2a + 1 + 2a - 1$
$= a^2$ ❹

EXAMINER'S TIP

Remember that $(a-1)^2$ means $(a-1)(a-1)$.

16 $x = 2.9$ ❹

EXAMINER'S TIP

When working out the solution of an equation by trial and improvement, you must show all steps in your working.

17 a) $-2, -1, 0, 1$ ❸
b) $p \leq 4.5$ ❷

EXAMINER'S TIP

Remember that solving inequalities is like solving equations.

18 a) $x < 7.5$ ❷
b) $-2 \leq x < 1.5$ ❷

19 $p = 3, q = -2$ ❹

20 $c = 3, d = 2$ ❹

EXAMINER'S TIP

Remember the coefficients of one of the letters need to be the same in both equations. In this example multiply the first equation by 3 and the second by 5.

21 a) $3(2a + 1)$ ❶
b) $4x(x + 2y)$ ❷
c) $(x + y)(x + y - 3)$ ❷
d) $\frac{5(x+2)}{(x+2)} = 5$ ❷
e) $(x - 6)(x + 6)$ ❷

EXAMINER'S TIP

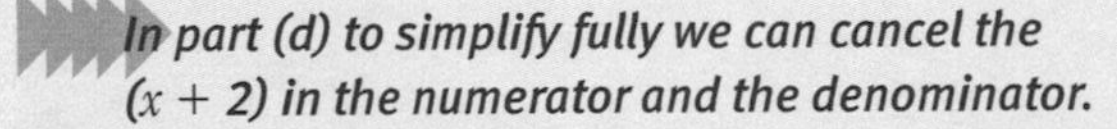

In part (d) to simplify fully we can cancel the $(x + 2)$ in the numerator and the denominator.

22 a) $(x - 2)(x - 3) = 0$
$x = 2, x = 3$ ❷
b) $(x + 3)(x - 4) = 0$
$x = -3, x = 4$ ❷

EXAMINER'S TIP

Factorise the equation first then solve.

23 a) $t = \frac{v - u}{a}$ ❷
b) $v = \pm\sqrt{\frac{2E}{m}}$ ❷

EXAMINER'S TIP

In part (b) isolate the v^2 before taking the square root, it could be a positive or negative.

24 a) 12 km/h ❷
b) Between 1100 and 1230 he was stationary. ❶
c)

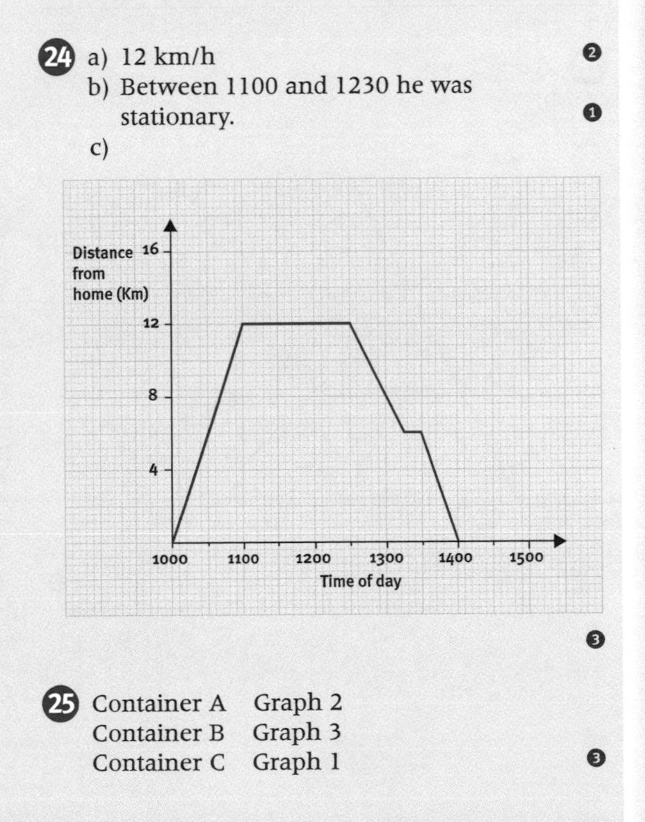

❸

25 Container A Graph 2
Container B Graph 3
Container C Graph 1 ❸

Algebra

26

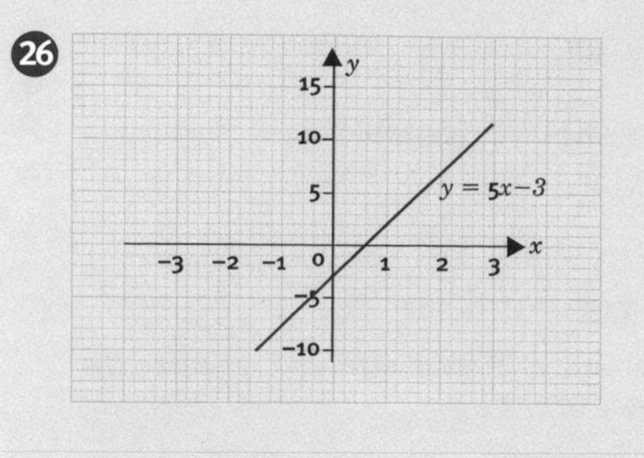

❸

EXAMINER'S TIP

Always try to choose suitable scales on the axes, when none are given.

27

A	$y = 5 - 2x$	✓
B	$y = 3x - 4$	
C	$2y + 4x = 7$	✓
D	$3y + y = 6$	
E	$6y - 2x = 5$	

❷

EXAMINER'S TIP

Rearrange the equation in the form '$y = mx+c$' before you compare them.

28 a) $x = 2, y = 3$ ❷

b)

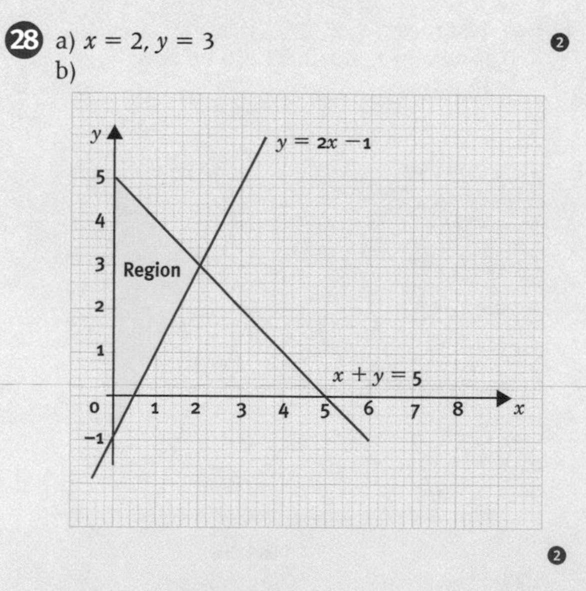

❷

29 a)

x	−3	−2	−1	0	1	2
$y = x^3 + 5$	−22	−3	4	5	6	13

❷

b) ❷

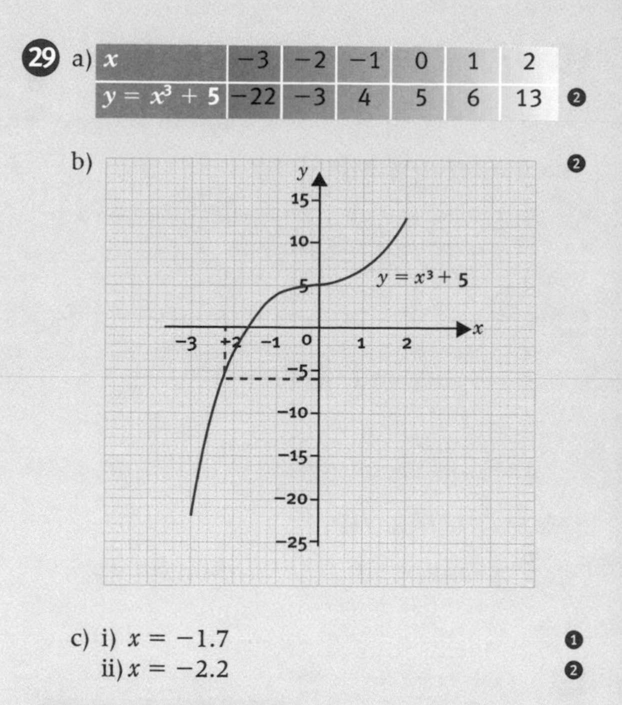

c) i) $x = -1.7$ ❶

ii) $x = -2.2$ ❷

EXAMINER'S TIP

When using the graph to find the solutions of the equation show your method lines.

30

Graph A	$xy = 4$
Graph B	$y = 3x - 4$
Graph C	$y = x^2 - 8$
Graph D	$y = 1 - 2x$

❹

FOR MORE INFORMATION ON THIS TOPIC ... SEE GCSE SUCCESS GUIDE ... 1 – PAGES 28–43

CHAPTER 3

Shape, space and measures

To revise this topic more thoroughly, see pages 46–71 in *Letts GCSE Success Guide*.

Try this sample GCSE question and then compare your answers with the Grade D and Grade B model answers on the next page.

1 The diagram shows the position of four towns.

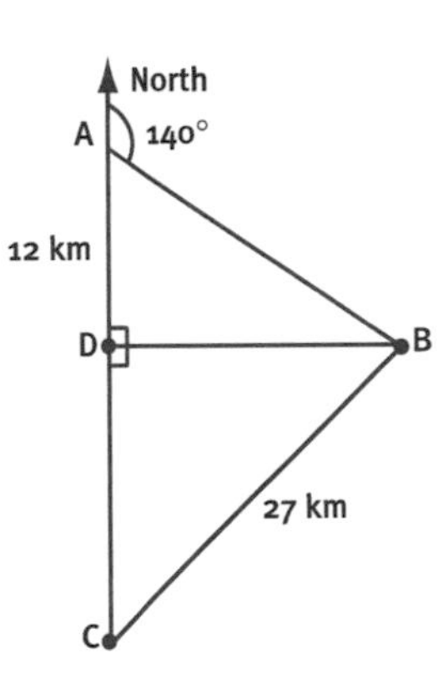

Town B is due east of town D.
Town C is 27 km from town B.
Town A is 12 km due north of town D.
Town B is on a bearing of 140° from town A.

a Calculate the distance from town D to town B. Give your answer in km, correct to 3 significant figures.

....

....

.... **[3]**

b Calculate the bearing of town B from town C, giving your answer to the nearest degree.

....

....

.... **[4]**

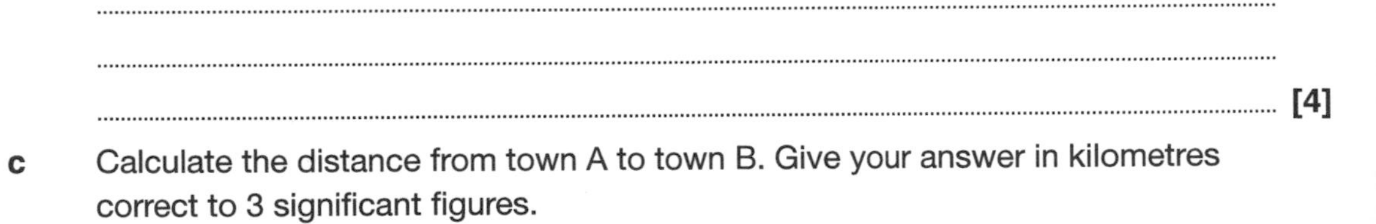

c Calculate the distance from town A to town B. Give your answer in kilometres correct to 3 significant figures.

....

.... **[3]**

Total 10 marks

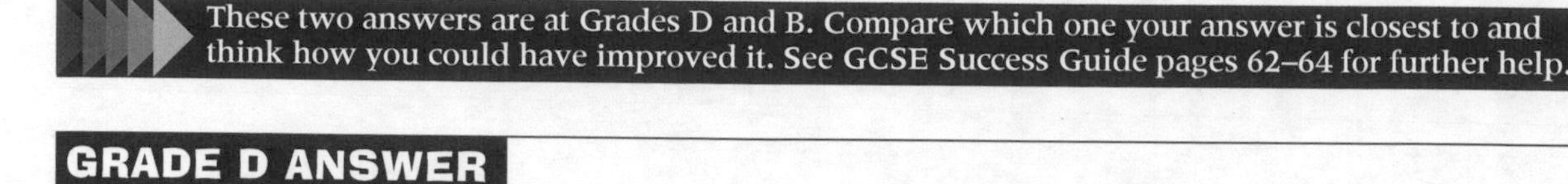

These two answers are at Grades D and B. Compare which one your answer is closest to and think how you could have improved it. See GCSE Success Guide pages 62–64 for further help.

GRADE D ANSWER

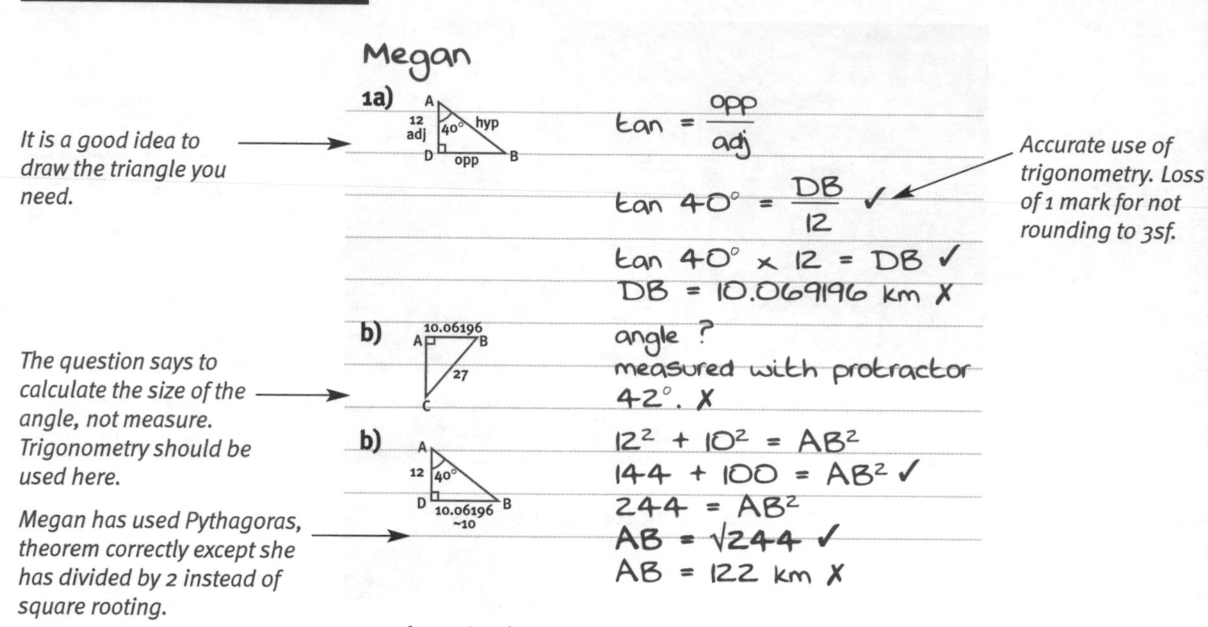

Megan

1a)

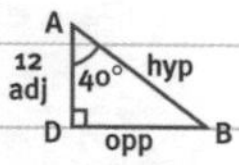

It is a good idea to draw the triangle you need.

$\tan = \dfrac{opp}{adj}$

$\tan 40° = \dfrac{DB}{12}$ ✓

Accurate use of trigonometry. Loss of 1 mark for not rounding to 3sf.

$\tan 40° \times 12 = DB$ ✓

$DB = 10.069196$ km ✗

b)

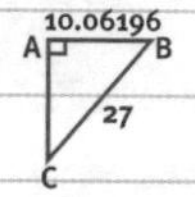

angle ?
measured with protractor
42°. ✗

The question says to calculate the size of the angle, not measure. Trigonometry should be used here.

b)

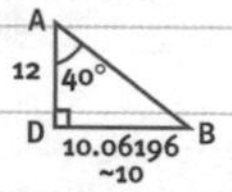

$12^2 + 10^2 = AB^2$

$144 + 100 = AB^2$ ✓

$244 = AB^2$

$AB = \sqrt{244}$ ✓

$AB = 122$ km ✗

Megan has used Pythagoras' theorem correctly except she has divided by 2 instead of square rooting.

4 marks = Grade D answer

Grade booster ····> move a D to a C

Megan has some understanding of trigonometry and Pythagoras' theorem. She needs to be more competent when using trigonometry to find angles in right angled triangles and more accurate when using Pythagoras' theorem.

GRADE B ANSWER

Richard

1a)

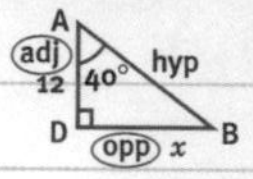

$\tan 40° = \dfrac{x}{12}$ ✓

$12 \times \tan 40° = x$ ✓

$x = 10.1$ km (3sf) ✓

Fully correct with well laid out working. Answer correctly given to 3sf.

b)

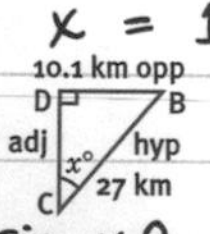

$\sin x° = \dfrac{10.0619\ldots}{27}$

It is good to see that Richard has not rounded off this value before finding the angle.

$\sin x = 0.37266\ldots$ ✓

$x = \sin^{-1}(0.37266\ldots)$ ✓

$= 21.88°$ ✓

$= 22°$ (nearest degree) ✗

A bearing must have three figures. The correct answer is therefore 022°.

c)

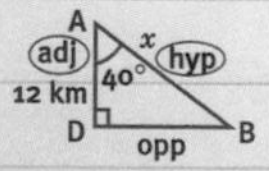

$\cos 40° = \dfrac{adj}{hyp}$ $\quad \cos 40° = \dfrac{12}{x}$ ✓

$x = \dfrac{12}{\cos 40°}$ ✓

$x = 15.7$ km (3sf) ✓

Richard has chosen to use trigonometry to answer this question. He could have used Pythagoras' theorem, which might have been slightly easier.

9 marks = Grade B answer

Richard has shown very good competence when using trigonometry. He should be in line for a grade B if his other work is of a similar standard.

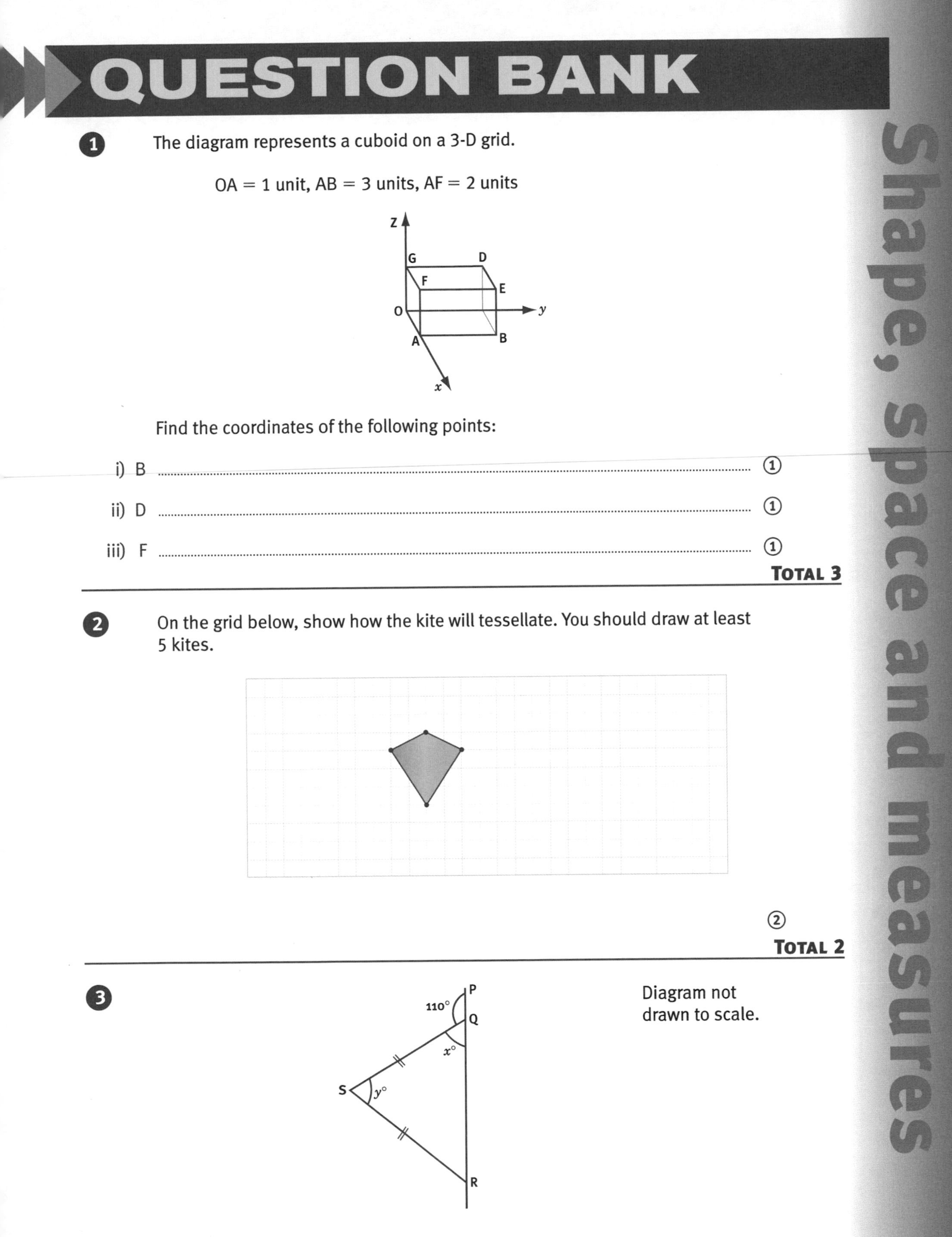

QUESTION BANK

1 The diagram represents a cuboid on a 3-D grid.

OA = 1 unit, AB = 3 units, AF = 2 units

Find the coordinates of the following points:

i) B .... ①

ii) D .... ①

iii) F .... ①

TOTAL 3

---

2 On the grid below, show how the kite will tessellate. You should draw at least 5 kites.

②

TOTAL 2

3

Diagram not drawn to scale.

PQR is a straight line.

ANSWERS ON PAGE 57 ANSWERS ON PAGE 57 ANSWERS ON PAGE 57 ANSWERS ON PAGE 57

a) i) Work out the size of the angle marked $x°$.

....

....° (1)

ii) Give a reason for your answer.

....

.... (1)

b) i) Work out the size of the angle marked $y°$.

....

....° (1)

ii) Give reasons for your answer.

....

....

.... (2)

TOTAL 5

4

135°
a
b

Diagram not drawn to scale.

a) Work out the size of angle a.

....

....° (1)

b) Give a reason for your answer.

....

....

.... (1)

c) Work out the size of angle b.

....

....° (1)

TOTAL 3

 ANSWERS ON PAGE 57 ANSWERS ON PAGE 57 ANSWERS ON PAGE 57 ANSWERS ON PAGE 57

5 The diagram shows a regular octagon.

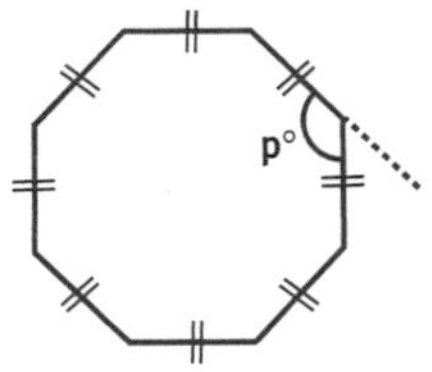

a) Work out the size of angle p.

....

....

....° (2)

b) The interior angle of a different regular polygon is 156°.

Work out the number of sides in this regular polygon.

....

....

.... (3)

TOTAL 5

6 The diagram shows the position of a boat (B) a lighthouse (L) and a buoy (Y).
The scale of the diagram is 1 cm to 5 km.

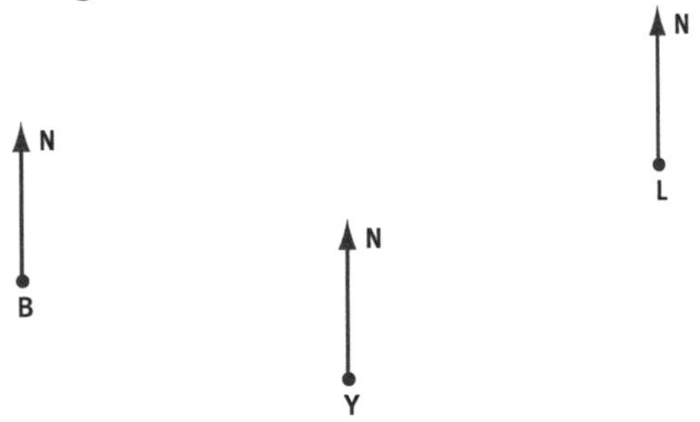

a) What is the distance of the boat from the lighthouse?
Measure the diagram accurately.

....

.... km (2)

b) Find the bearing of

i) B from Y

....° (1)

ii) L from Y

....° (1)

TOTAL 4

ANSWERS ON PAGE 57 ANSWERS ON PAGE 57 ANSWERS ON PAGE 57 ANSWERS ON PAGE 57

Shape, space and measures

7

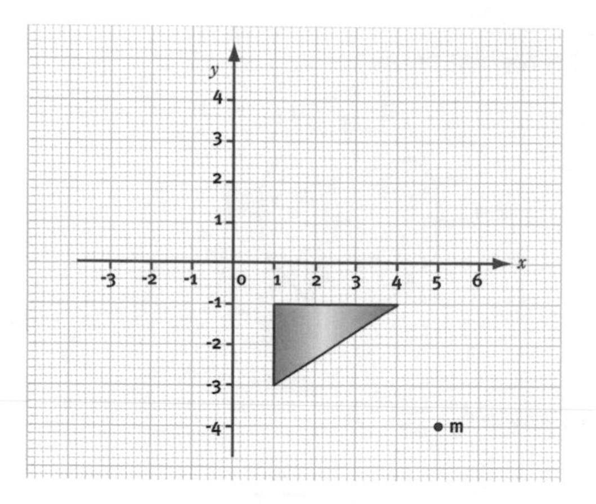

a) Rotate the shaded triangle 90° anticlockwise about O. Label P. ③

b) Enlarge the shaded triangle with centre **m** and scale factor 2. Label R. ③

TOTAL 6

8

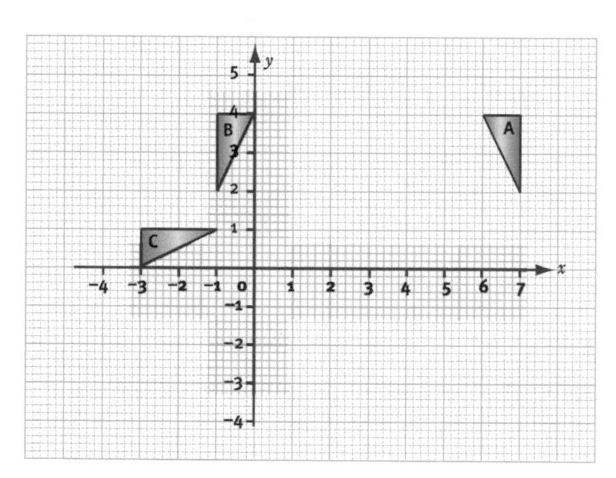

Triangle A is a reflection of triangle B.

a) Write down the equation of the mirror line.

....

.... ①

b) Translate triangle C by the vector $\binom{5}{-3}$. Call the new triangle D. ②

TOTAL 3

---

9 a) Fiona weighs 46 kg, to the nearest kg.
Between what limits does Fiona's weight lie?

....

.... ②

b) Saimia ran the 10 m race in a time of 14.3 seconds to the nearest tenth of a second. What is the shortest possible time in which she ran the race?

....

.... ②

TOTAL 4

ANSWERS ON PAGE 57 ANSWERS ON PAGE 57 ANSWERS ON PAGE 57 ANSWERS ON PAGE 57

10 There are 14 pounds in a stone.
There are 2.2 pounds in a kilogram.
A woman weighs 8 stone 6 pounds.
Work out her weight in kilograms.
Give your answer correct to the nearest kilogram.

....

....

....

.... kg ③

TOTAL 3

---

11 a) Jonathan drives at an average speed of 65 mph for 2 hours and 20 minutes.
How far has Jonathan travelled?

....

.... miles ②

b) Mr Wong travels at an average speed of 62 mph.
His journey is 125 miles. How long does his journey take?

....

.... ②

TOTAL 4

12

4 cm
3 cm
5 cm

a) Work out the area of the triangle.

....

.... cm^2 ②

The triangle is enlarged by a scale factor 3.

b) Work out the area of the enlarged triangle.

....

....

.... cm^2 ②

TOTAL 4

Shape, space and measures

ANSWERS ON PAGE 57 ANSWERS ON PAGE 57 ANSWERS ON PAGE 57 ANSWERS ON PAGE 57

13 A circle is cut out from a piece of rectangular wood.

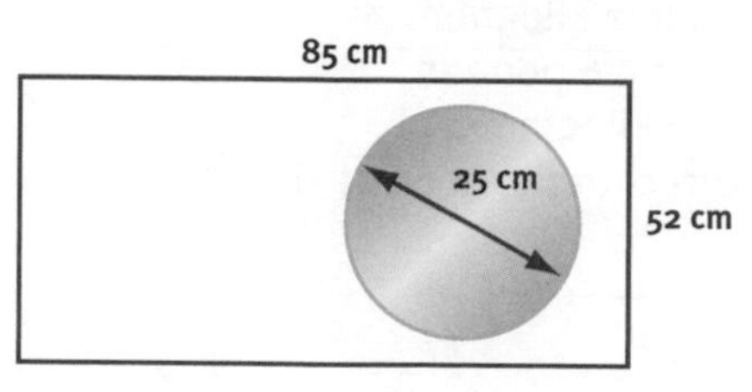

Diagram not drawn to scale.

The length and width of the rectangle are 85 cm and 52 cm.
The circle has a diameter 25 cm.

Calculate the area of the piece of wood that remains once the circle has been removed. Give the units of your answer.

(5)

TOTAL 5

14

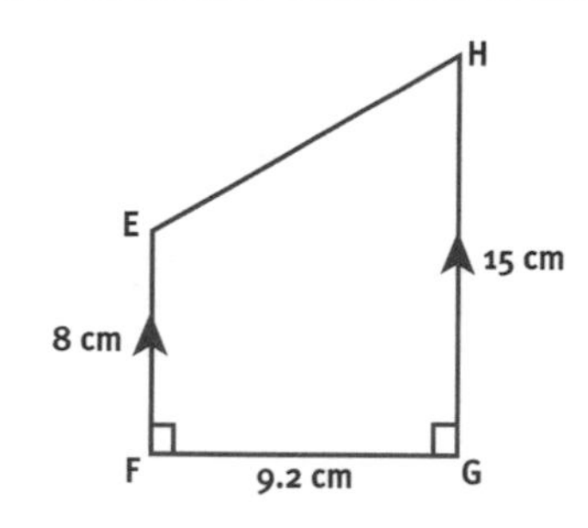

The diagram shows a trapezium.
EF is parallel to GH.
EF = 8 cm, FG = 9.2 cm, GH = 15 cm.

a) Calculate the area of trapezium EFGH.

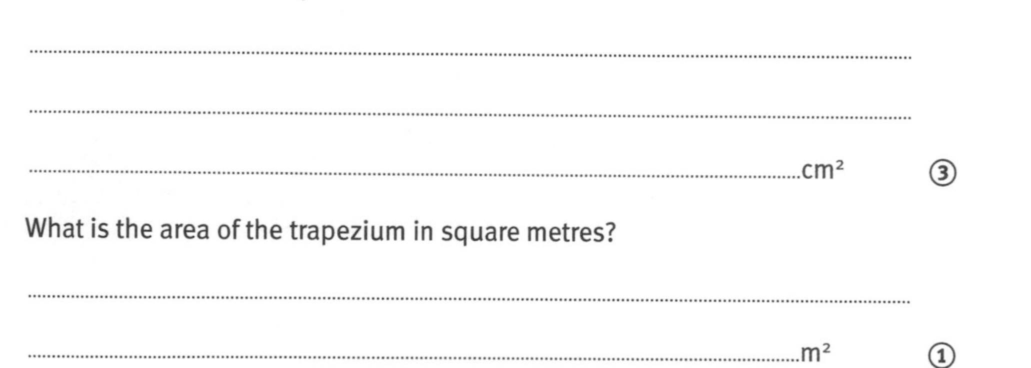

.... cm^2 (3)

b) What is the area of the trapezium in square metres?

.... m^2 (1)

TOTAL 4

15 A solid cuboid has a volume of 288 cm³. The length of the cuboid is 8 cm and the width is 4 cm. Work out the total surface area of the cuboid.

....

....

....

....

....

.... cm² (4)

TOTAL 4

---

16 The diagram shows a solid cylinder A.
The radius of the cylinder is 6 cm.
The height of the cylinder is 14 cm.

6 cm
A
14 cm

a) Calculate the volume of the cylinder.
Give your answer correct to 3 significant figures.

....

....

....

....

.... cm³ (2)

A second cylinder has the same volume as cylinder A.
The height of the cylinder is 9 cm.

9 cm

b) Calculate the radius of the second cylinder.
Give your answer correct to 1 decimal place.

....

....

....

....

.... cm (3)

TOTAL 5

ANSWERS ON PAGE 57 ANSWERS ON PAGE 57 ANSWERS ON PAGE 57 ANSWERS ON PAGE 57

17 The table shows six expressions.
p, q and r are lengths.
π, 2 and 5 are numbers and have no dimensions.

$5pqr$	$p(q + r)$	$\pi p + 5r$	$2\pi p(q^2 + r^2)$	$\pi p^2 - 5qr$	$\sqrt{2q^2 + r^2}$

i) Put the letter P in the box underneath each of the **two** expressions that could represent a **perimeter**.

ii) Put the letter V in the box underneath each of the **two** expressions that could represent a **volume**.

.... ③

TOTAL 3

18

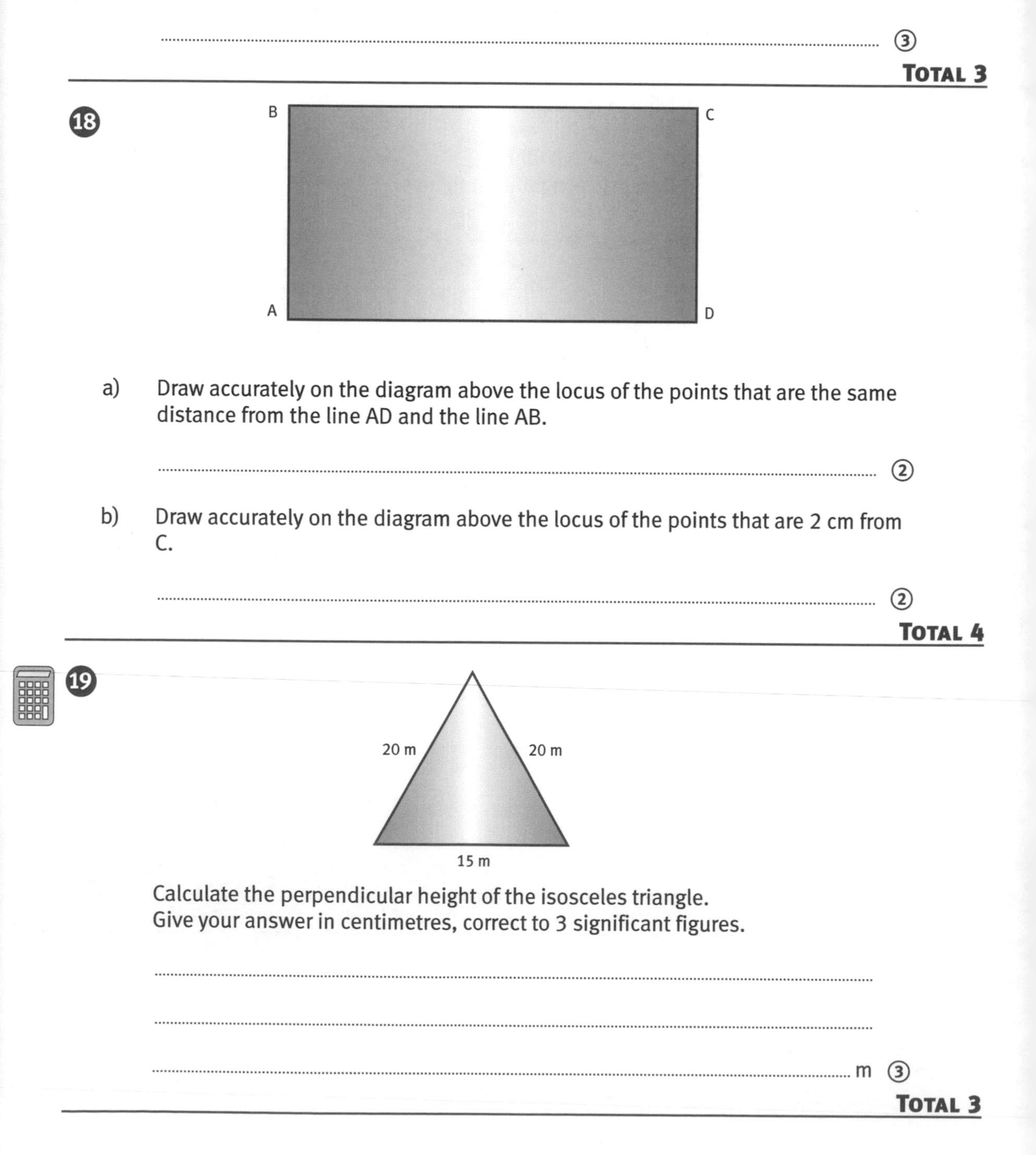

a) Draw accurately on the diagram above the locus of the points that are the same distance from the line AD and the line AB.

.... ②

b) Draw accurately on the diagram above the locus of the points that are 2 cm from C.

.... ②

TOTAL 4

19

Calculate the perpendicular height of the isosceles triangle.
Give your answer in centimetres, correct to 3 significant figures.

....

....

.... m ③

TOTAL 3

20 A ladder of length 5 metres rests so that the foot of the ladder is 2.7 metres away from a wall. Calculate how far up the wall the ladder reaches. Give your answer to 2 significant figures.

....

....

.... m ③

TOTAL 3

---

21 a) Calculate the length of PQ in this diagram.

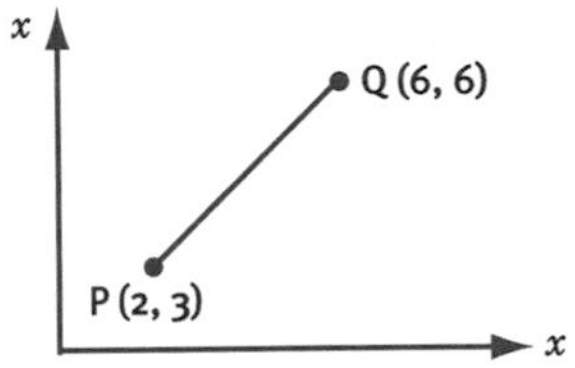

....

.... ②

b) Calculate the coordinates of the midpoint of PQ.

....

.... ②

TOTAL 4

22

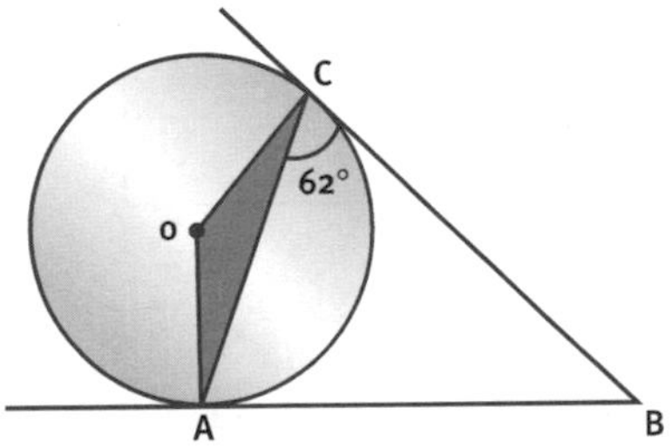

A and C are points on the circumference of a circle, centre O.
AB and BC are tangents to the circle.
Angle BCA = 62°

a) Explain why angle ACO = 28°.

....

.... ①

b) Calculate the size of angle AOC.
Give reasons for your answer.

....

.... ③

TOTAL 4

Shape, space and measures

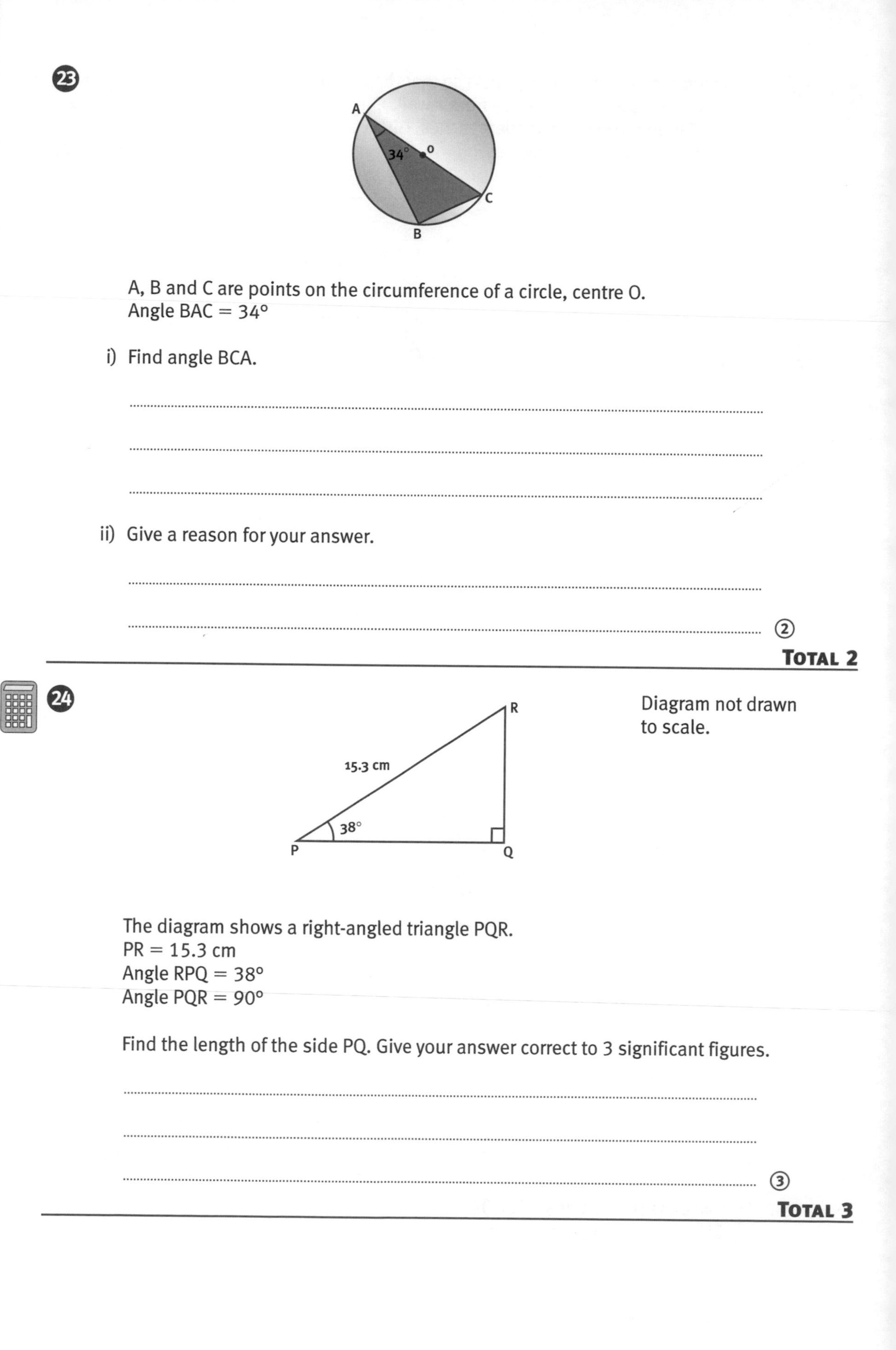

23

A, B and C are points on the circumference of a circle, centre O.
Angle BAC = 34°

i) Find angle BCA.

....

....

....

ii) Give a reason for your answer.

....

.... ②

TOTAL 2

24

Diagram not drawn to scale.

The diagram shows a right-angled triangle PQR.
PR = 15.3 cm
Angle RPQ = 38°
Angle PQR = 90°

Find the length of the side PQ. Give your answer correct to 3 significant figures.

....

....

.... ③

TOTAL 3

ANSWERS ON PAGE 57 ANSWERS ON PAGE 57 ANSWERS ON PAGE 57 ANSWERS ON PAGE 57

25 The diagram shows the positions of three towns A, B and C.

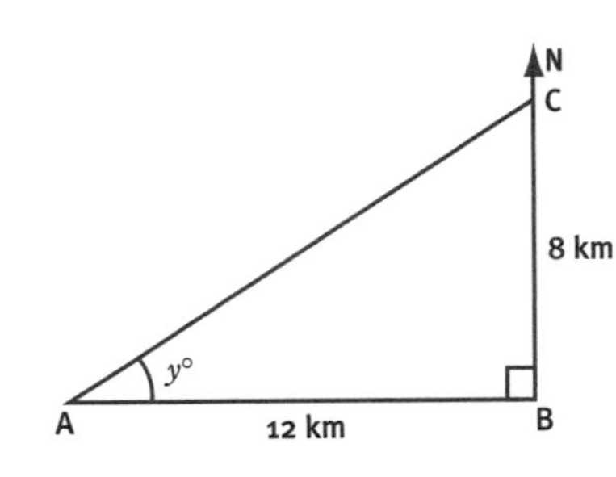

Town A is 12 km due west of town B.
Town C is 8 km due north of town B.

a) Calculate the size of the angle marked $y°$.
Give your answer correct to 1 decimal place.

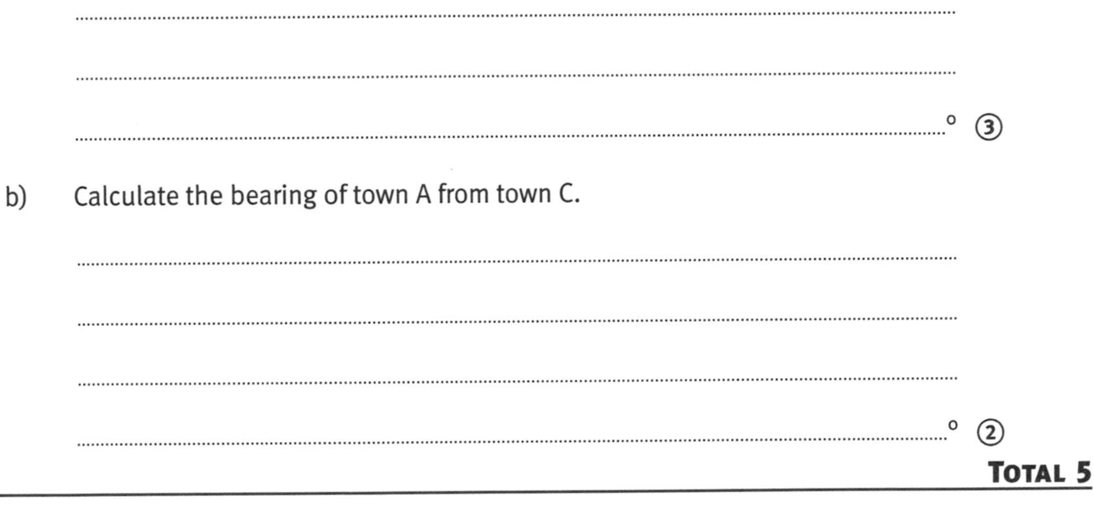

....

....

....

....° ③

b) Calculate the bearing of town A from town C.

....

....

....

....° ②

TOTAL 5

26 The diagram shows a right-angled triangle ABC.

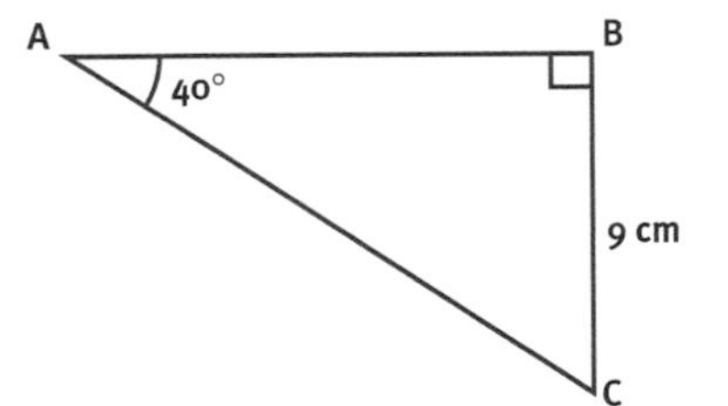

BC is 9 cm.
Angle ABC = 90°
Angle CAB = 40°

Find the length of the side AB. Give your answer correct to 3 significant figures.

....

....

.... cm ③

TOTAL 3

ANSWERS ON PAGE 57 ANSWERS ON PAGE 57 ANSWERS ON PAGE 57 ANSWERS ON PAGE 57

27 The two cylinders are similar. Calculate the height (x) of the larger cylinder. Give your answer to 1 decimal place.

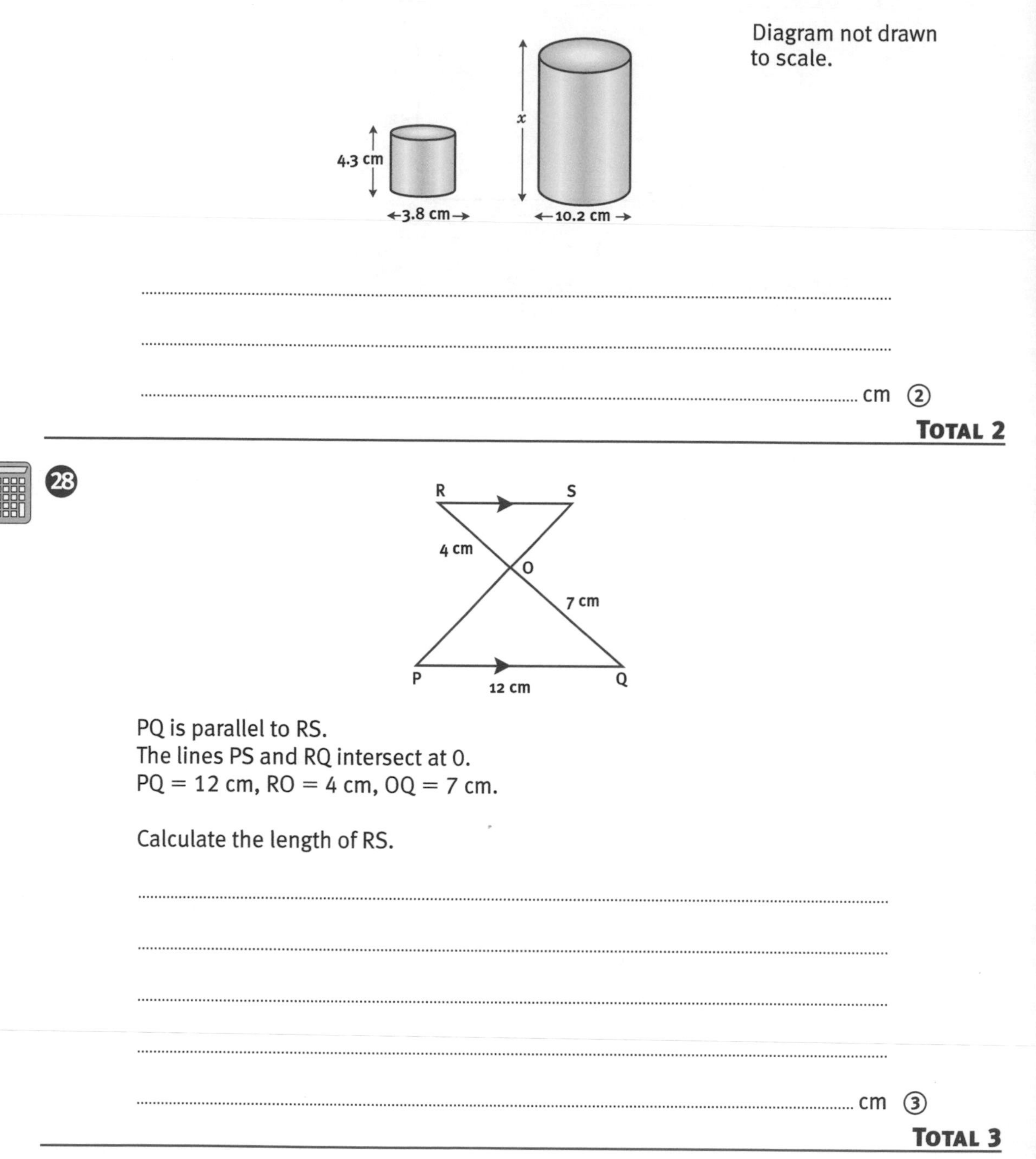

....

....

.... cm (2)

TOTAL 2

28

PQ is parallel to RS.
The lines PS and RQ intersect at O.
PQ = 12 cm, RO = 4 cm, OQ = 7 cm.

Calculate the length of RS.

....

....

....

....

.... cm (3)

TOTAL 3

QUESTION BANK ANSWERS

1

i) (1, 3, 0) (1)
ii) (0, 3, 2) (1)
iii) (1, 0, 2) (1)

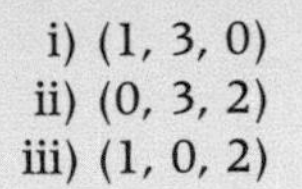

EXAMINER'S TIP

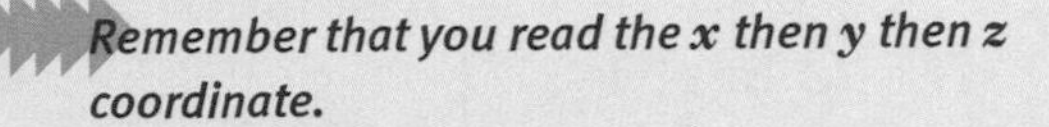

Remember that you read the x then y then z coordinate.

2

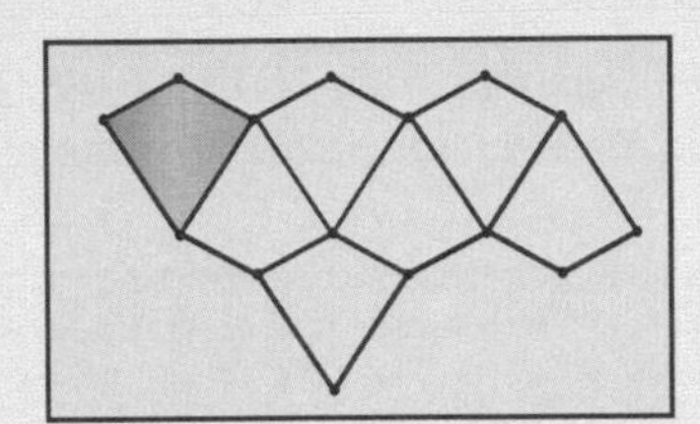

(2)

EXAMINER'S TIP

Remember to draw at least five kites, otherwise you will lose marks.

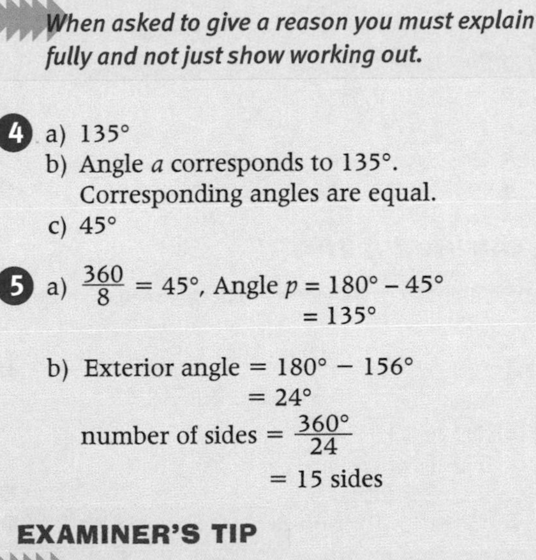

3 a) i) 70° (1)
ii) Angles on a straight line add up to 180° (1)
b) i) 40° (1)
ii) Triangle SQR is an isosceles triangle, hence angle SRQ is also 70°. Angles in a triangle add up to 180°, hence angle y is 40°. (2)

EXAMINER'S TIP

When asked to give a reason you must explain fully and not just show working out.

4 a) 135° (1)
b) Angle a corresponds to 135°. Corresponding angles are equal. (1)
c) 45° (1)

5 a) $\frac{360}{8} = 45°$, Angle $p = 180° - 45° = 135°$ (2)

b) Exterior angle $= 180° - 156° = 24°$

number of sides $= \frac{360°}{24} = 15$ sides (3)

EXAMINER'S TIP

Part (b) is a multistep question. You need to work out the size of the exterior angle first.

6 a) 6.6 cm
6.6×5 km
$= 33$ km (2)
b) i) 286° (1)
ii) 057° (1)

EXAMINER'S TIP

Bearings always have 3 figures so it must be 057° in (b) (ii).

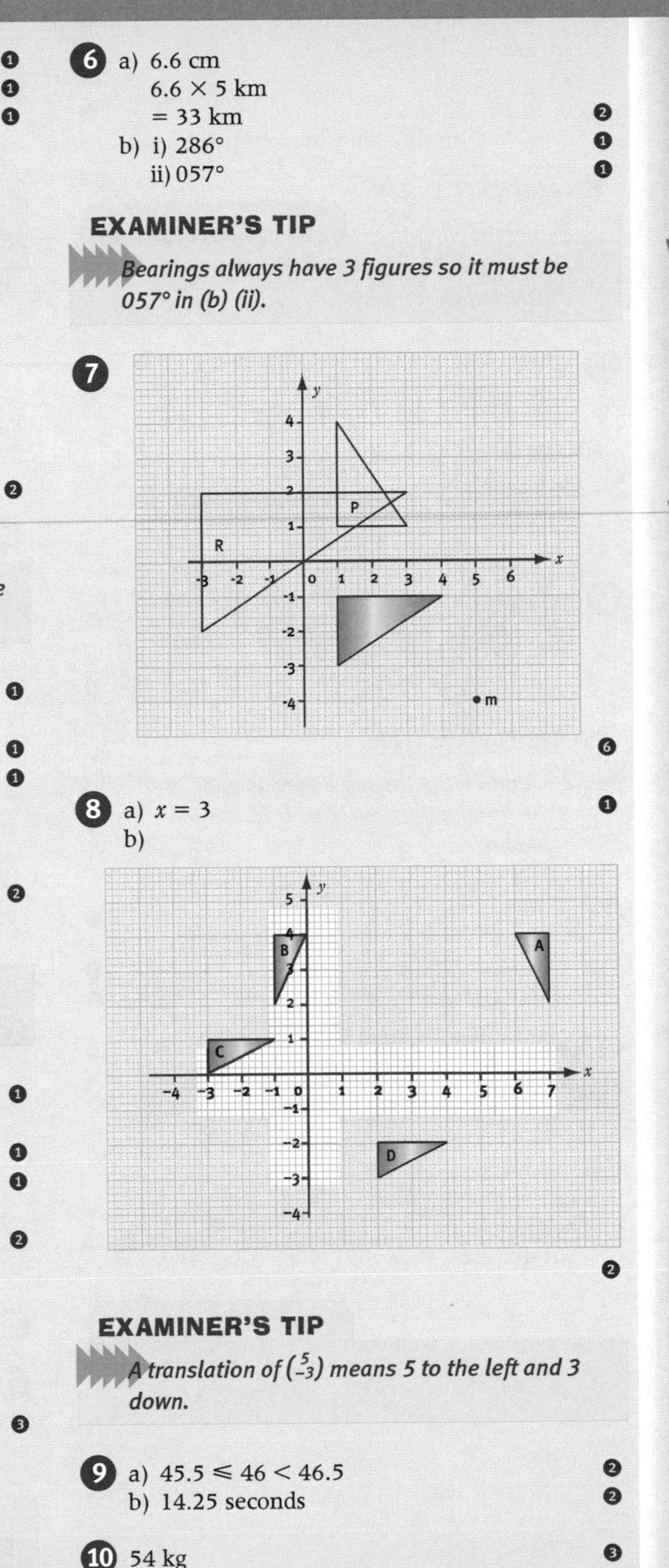

7

(6)

8 a) $x = 3$ (1)
b)

(2)

EXAMINER'S TIP

A translation of $\binom{5}{-3}$ means 5 to the left and 3 down.

9 a) $45.5 \leqslant 46 < 46.5$ (2)
b) 14.25 seconds (2)

10 54 kg (3)

Shape, space and measures

11 a) distance = speed × time
$= 65 \times 2^{20}/_{60}$
$= 151.\dot{6}$ miles
= 152 miles (nearest mile) (2)

b) $s = \frac{d}{t}$, $t = \frac{d}{s}$, $t = \frac{125}{62}$

$t = 2.016 \ldots$ (2)
= 2 hours 1 min (nearest minute)

EXAMINER'S TIP

Remember that there are 60 minutes in 1 hour. For part (a), 20 minutes needs to be written as a fraction of 60 minutes.

12 a) 6 cm² (2)
b) 54 cm² (2)

EXAMINER'S TIP

The area of the enlarged shape will be 3^2 i.e. 9 times bigger.

13 Rectangle area : $85 \times 52 = 4420$ cm²
Circle area: $\pi \times 12.5^2 = 490.87$ cm²
Area left : $4420 - 490.87$
$= 3929.1$ cm² (5)

EXAMINER'S TIP

Remember that the radius needs to be used when finding the area of a circle, not the diameter.

14 a) $\frac{(8 + 15)}{2} \times 9.2$ (3)

Area = 105.8 cm² (3)
b) 0.01058 m² (1)

EXAMINER'S TIP

Remember that 1 m² = 10 000 cm².

15 $V = lwh$
$288 = 8 \times 4 \times h$
height = 9 cm
Surface area $= (9 \times 4) \times 2 + (8 \times 4) \times 2 + (8 \times 9) \times 2$
$= 72 + 64 + 144$
$= 280$ cm² (4)

16 a) $V = \pi r^2 h$
$= \pi \times 6^2 \times 14$
$= 1583.4$ cm³
$= 1580$ cm³ (3sf) (2)

b) $1583.4 = \pi r^2 h$
$1583.4 = \pi \times r^2 \times 9$
$r = \sqrt{\frac{1583.4}{9\pi}}$
$r = \sqrt{56}$
$r = 7.48$ cm
$r = 7.5$ cm (1 dp) (3)

EXAMINER'S TIP

In part (b) it is sometimes easier to substitute the values in the formula and then rearrange.

17

$5pqr$	$p(q + r)$	$\pi p + 5r$	$2\pi p\,(q^2 + r^2)$	$\pi p^2 - 5qr$	$\sqrt{2q^2 + r^2}$
V		P	V		P

(3)

18

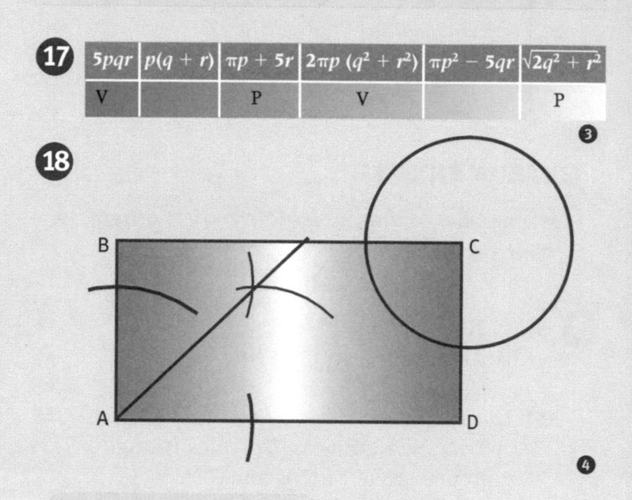

(4)

EXAMINER'S TIP

For part (a) use a pair of compasses to bisect the angle at A.

19 $20^2 = h^2 + 7.5^2$
$h^2 = 20^2 - 7.5^2$
$h^2 = 343.75$
$h = \sqrt{343.75}$
$h = 1850$ cm (3)

EXAMINER'S TIP

Remember to halve the base distance.

20 4.2 m (3)

21 a) $PQ = \sqrt{3^2 + 4^2}$
$PQ = \sqrt{25}$
PQ = 5 units (2)
b) $\frac{1}{2}(8)$, $\frac{1}{2}(9)$ Coordinates are (4, 4.5) (2)

EXAMINER'S TIP

Use Pythagoras' theorem to calculate the length of PQ.

FOR MORE INFORMATION ON THIS TOPIC … SEE GCSE SUCCESS GUIDES … PAGES 46–71

22 a) Angle BCO is 90°,
(tangent and radius meet at 90°)
Angle ACO is 90° − 62° = 28° (1)

b) Angle ACO = 28°
Angle CAO = 28° since triangle AOC is isosceles.
Angle AOC is 180° − 56°
(angles in a triangle)
AOC = 124° (3)

23 i) 56° (1)

ii) Triangle ABC is right-angled at B, as the triangle is in a semicircle.
Angles in a triangle add up to 180°, hence angle BCA = 56° (1)

EXAMINER'S TIP

Make explanations clear and explain with angle properties not just by showing working out.

24 12.1 cm (3)

25 a) $\tan y = \frac{8}{12}$ (3)

$y = 33.7°$

b) 236.3° (2)

EXAMINER'S TIP

Label the sides of the triangle before you decide on which ratio to use.

26 $\tan 40° = \frac{9}{AB}$

$AB = \frac{9}{\tan 40°}$

AB = 10.73 cm
= 10.7 cm (3sf) (3)

EXAMINER'S TIP

Take extra care on a question like this when the unknown is on the denominator.

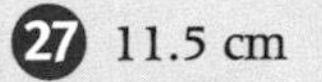

27 11.5 cm (2)

28 $\frac{RS}{4} = \frac{12}{7}$

$RS = \frac{12}{7} \times 4$

RS = 6.9 cm (3)

EXAMINER'S TIP

Always check that the answer sounds sensible.

CHAPTER 4

Handling data

To revise this topic more thoroughly, see pages 74–89 in *Letts GCSE Success Guide.*

Try this sample GCSE question and then compare your answers with the Grade D and Grade B model answers on the next page.

1 The times taken by 50 people to complete an IQ test are shown, in minutes, in the table below.

Time (t mins)	Frequency
$20 \leq t < 30$	2
$30 \leq t < 40$	10
$40 \leq t < 50$	21
$50 \leq t < 60$	10
$60 \leq t < 70$	6
$70 \leq t < 80$	1

a Calculate an estimate of the mean time taken.

.... **[4]**

b Complete the cumulative frequency table and draw the cumulative frequency graph of the distribution.

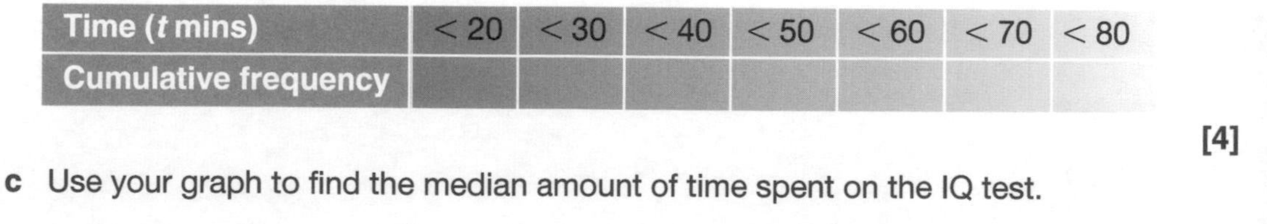

Time (t mins)	< 20	< 30	< 40	< 50	< 60	< 70	< 80
Cumulative frequency							

[4]

c Use your graph to find the median amount of time spent on the IQ test.

....

.... **[1]**

d How many people took more than 62 minutes to complete the test?

....

....

.... **[2]**

Total 11 marks

These two answers are at Grades D and B. Compare which one your answer is closest to and think how you could have improved it. See GCSE Success Guide page 84 for further help.

GRADE D ANSWER

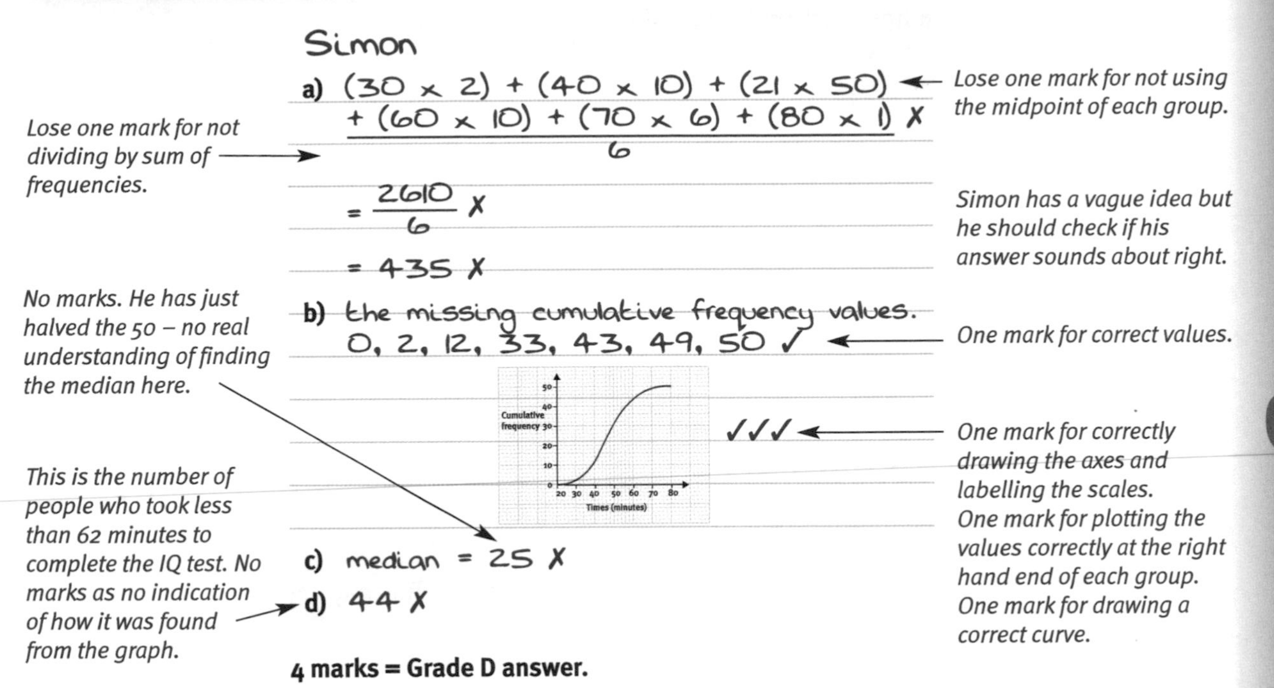

Simon

a) (30 x 2) + (40 x 10) + (21 x 50) + (60 x 10) + (70 x 6) + (80 x 1) / 6 ✗

= 2610 / 6 ✗

= 435 ✗

Lose one mark for not using the midpoint of each group.

Lose one mark for not dividing by sum of frequencies.

Simon has a vague idea but he should check if his answer sounds about right.

b) the missing cumulative frequency values.
0, 2, 12, 33, 43, 49, 50 ✓

One mark for correct values.

✓✓✓

One mark for correctly drawing the axes and labelling the scales. One mark for plotting the values correctly at the right hand end of each group. One mark for drawing a correct curve.

c) median = 25 ✗

No marks. He has just halved the 50 – no real understanding of finding the median here.

d) 44 ✗

This is the number of people who took less than 62 minutes to complete the IQ test. No marks as no indication of how it was found from the graph.

4 marks = Grade D answer.

Grade booster ····> move a D to a C

Simon needs to revise his understanding of how to calculate an estimate of the mean for grouped data. He also needs to be sure he knows how to find the median and quartiles and practise reading values on scales.

GRADE B ANSWER

Rebecca

a) (25 x 2) + (35 x 10) + (45 x 21) + (55 x 10) + (65 x 6) + (75 x 1) / 50 ✓✓

= 2360 / 50 ✓

= 47.2 mins. ✓

Two marks for correctly using the midpoints to work out an estimate of the total. One mark for dividing by 50.

b) the missing cumulative frequency values are
0, 2, 12, 33, 43, 49, 50 ✓

Correct values given.

✓✓✓

One mark for correctly drawing axes and labelling scales. One mark for plotting the values correctly at the right hand end of each group. One mark for drawing a correct curve.

c) median approx 46 mins. ✓

Correct answer given and clear indication from the graph of how she obtained the answer.

d) approx 44. ✗

This is the number of people who took less than 62 minutes to complete the IQ test. No marks as there is no indication of how it was found from the graph.

9 marks = Grade B answer

Rebecca has a good understanding of calculating an estimate of the mean for grouped data. She is competent at drawing and finding the median from cumulative frequency graphs.

Handling data

QUESTION BANK

1 Edward is carrying out a survey into the types of films his friends like to watch. Draw a suitable data collection sheet that Edward could use.

....

....

....

....

.... ③

TOTAL 3

---

2 Charlotte is carrying out a survey into how much homework pupils in Year 8 receive on a Tuesday night.

She asked this question:

'How many hours homework do you usually get on a Tuesday night?'

Reece is also doing the same survey. His question says:

'On a Tuesday night, approximately how much homework do you get, to the nearest 30 minutes?'

Please tick a box:

0–30 mins ☐ 31–60 mins ☐ 61–90 mins ☐ over 90 mins ☐

Which student has the best question. Give a reason for your answer.

....

....

....

.... ③

TOTAL 3

ANSWERS ON PAGE 78 ANSWERS ON PAGE 78 ANSWERS ON PAGE 78 ANSWERS ON PAGE 78

80 pupils were asked about the type of books they were reading. The table gives some information about the answers.

	Fiction	Non-fiction	No books	Total
Boys	18		4	40
Girls		5		
Total	43			80

a) Complete the table. 

b) How many girls were reading no books?

.... ①

TOTAL 3

4 Some pupils were asked about the languages they are studying for GCSE. The table gives some information about their answers.

	French	German	Spanish	Total
Year 10	38	42		
Year 11	63		21	135
Total				250

a) Complete the table. ②

b) How many students are studying German for their GCSE?

....

....

....

.... ①

TOTAL 3

5 Joshua carried out a survey of students' favourite colours.
The table shows his results.

Colour	Frequency
Red	18
Blue	12
Green	16
Yellow	8

Complete an accurate pie chart to show this information.
Use the circle given below.

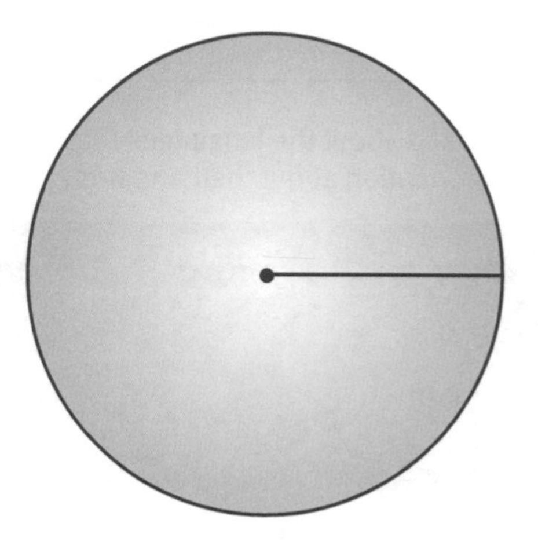

(4)

TOTAL 4

6 The pie chart gives information about how Imran spends a typical day.

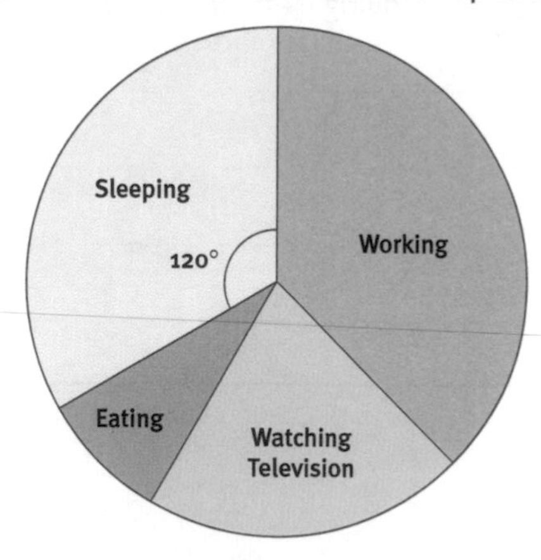

a) For how many hours does Imran work?

.... (2)

b) For how many hours does Imran watch television?

.... (2)

TOTAL 4

ANSWERS ON PAGE 78 ANSWERS ON PAGE 78 ANSWERS ON PAGE 78 ANSWERS ON PAGE 78

7 The number of millimetres of rainfall that fell during the first eight days of June are shown in the table below.

Day	1	2	3	4	5	6	7	8
Rainfall (mm)	8	10	4	2	2	6	7	10

a) Plot these figures on a graph. Use a scale of 1 cm for each day on the horizontal axis and 1 cm for every 2 mm of rain on the vertical axis.

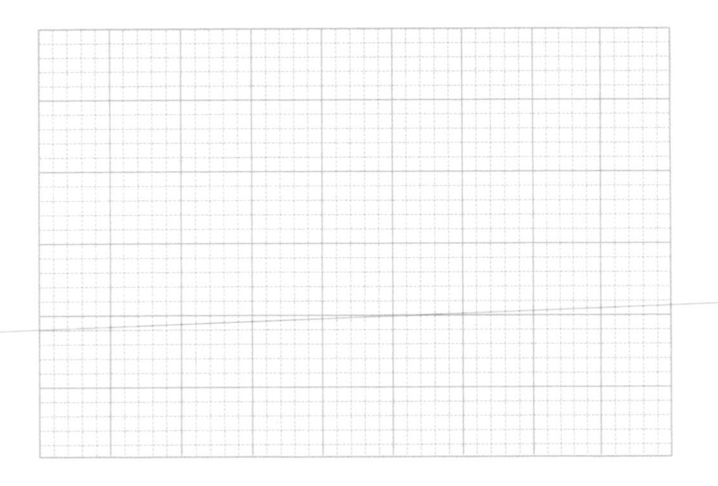

④

b) Complete the table below, showing the four-point moving average.

Day	mm of Rainfall	Moving Average
1	8	
2	10	6
3	4	4.5
4	2	.....
5	2	.....
6	6	.....
7	7	
8	10	

③

c) Plot the four-point moving average on the graph. ③

d) Comment on the general trend.

....

....

.... ②

TOTAL 12

ANSWERS ON PAGE 78 ANSWERS ON PAGE 78 ANSWERS ON PAGE 78 ANSWERS ON PAGE 78

8 Some students are doing a problem-solving test. The times, to the nearest minute, for the students to complete the test are recorded below.

Here are the results:

34	23	18	42	36	25	23	17	26	57
36	12	27	41	27	52	38	49	33	17

Draw a stem and leaf diagram to show this information.

(4)

TOTAL 4

9 The table shows the heights of a class of students.

Height in cm	Frequency
$130 \leqslant h < 135$	5
$135 \leqslant h < 140$	10
$140 \leqslant h < 145$	7
$145 \leqslant h < 150$	6
$150 \leqslant h < 155$	2
$155 \leqslant h < 160$	1

Draw a frequency polygon for this distribution.

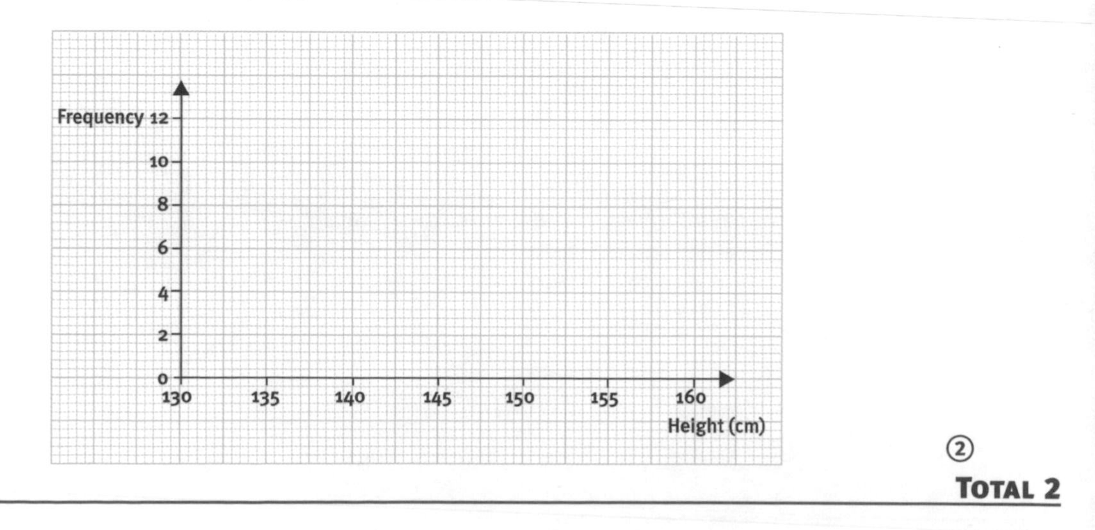

(2)

TOTAL 2

 ANSWERS ON PAGE 78 ANSWERS ON PAGE 78 ANSWERS ON PAGE 78 ANSWERS ON PAGE 78

10 The table shows the hours of sunshine and the number of ice creams sold in a café, during 10 days last summer.

Sunshine (hours)	5	7	4	2	9	9	8	6	3	4
Number of ice creams sold	19	24	16	9	32	29	26	23	13	18

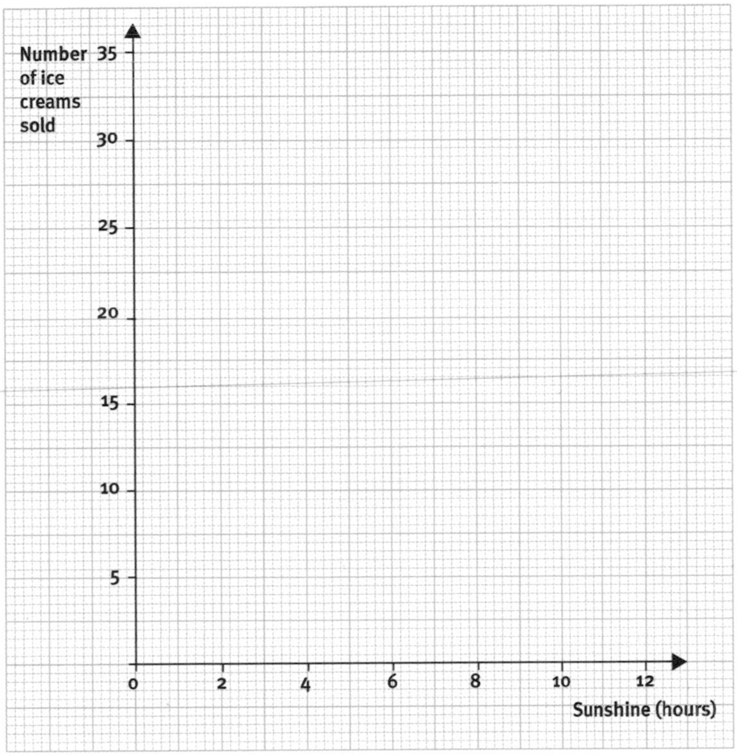

a) Draw a scatter diagram of the data given in the table above. (2)

b) Describe the relationship between the hours of sunshine and the number of ice creams sold.

....

.... (1)

c) Draw a line of best fit on your scatter diagram. (1)

d) Use your line of best fit to estimate:

i) the number of ice creams sold when there are 10 hours of sunshine.

.... (1)

ii) the amount of sunshine when there are 22 ice creams sold.

.... (1)

TOTAL 6

11 In a survey the heights of eight girls and their shoe sizes were measured.

Height in cm	150	157	159	161	158	164	152	168
Shoe size	3	5	$5^1/_2$	6	5	$6^1/_2$	$3^1/_2$	7

a) On the grid below, draw a scatter diagram to show this information.

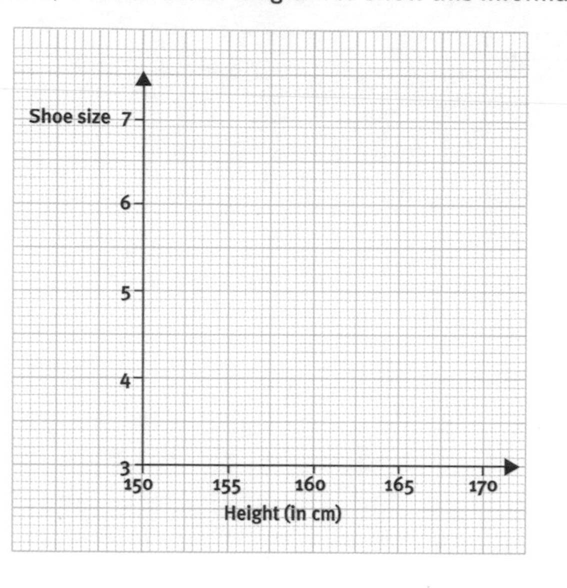

(4)

b) What type of correlation is there between height and shoe size?

....

.... (1)

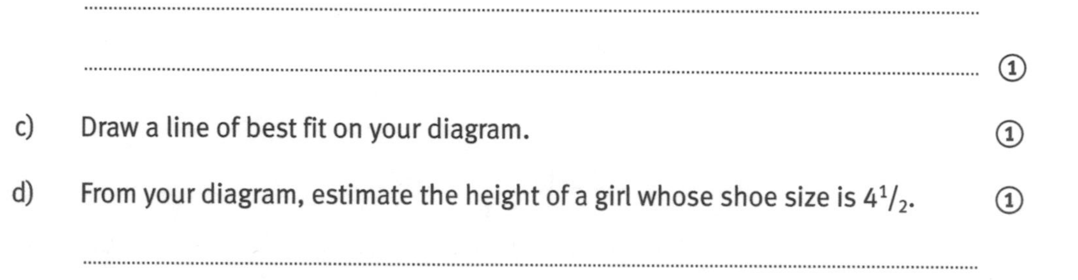

c) Draw a line of best fit on your diagram. (1)

d) From your diagram, estimate the height of a girl whose shoe size is $4^1/_2$. (1)

....

TOTAL 7

12 In a school, teachers can award merits for good work by students. Some teachers were asked how many merits they gave out last week. The results are shown in the table.

Number of merits	Number of teachers
0	2
1	5
2	9
3	10
4	8
5	4
6	1

ANSWERS ON PAGE 78 ANSWERS ON PAGE 78 ANSWERS ON PAGE 78 ANSWERS ON PAGE 78

a) Which number of merits is the mode?

.... ①

b) Find the median number of merits.

....

....

.... ②

c) Find the mean number of merits given out by the teachers.

....

....

....

.... ③

TOTAL 6

13 A die is thrown and the scores are noted. The results are shown in the table below.

Die score	Frequency
1	12
2	15
3	10
4	8
5	14
6	13

Work out the mean die score.

....

....

....

.... ③

TOTAL 3

14 The mean of 12, 5, 8 and x is 8. Work out the value of x.

....

....

.... ②

TOTAL 2

15 Some students do a test. There are 15 girls and 17 boys in the class.
In the test, the mean mark for the boys is 65.
In the test, the mean mark for the girls is 69.

What is the mean mark for the whole class of 32 students?

....

....

....

.... ③

TOTAL 3

16 Katie asked 150 people how much they spent last year on CDs.
The results are given in the table below:

Amount (£x)	Frequency
$0 < x \leq 20$	4
$20 < x \leq 40$	35
$40 < x \leq 60$	42
$60 < x \leq 80$	59
$80 < x \leq 100$	10

a) Calculate an estimate of the mean amount spent on CDs.

....

....

....

.... ④

b) Explain briefly why this value of the mean is only an estimate.

....

....

.... ①

TOTAL 5

ANSWERS ON PAGE 78 ANSWERS ON PAGE 78 ANSWERS ON PAGE 78 ANSWERS ON PAGE 78

17 The weights of some students in a class are measured. These are the results:

Weight in kg	Number of students
$40 \leqslant W < 45$	7
$45 \leqslant W < 50$	9
$50 \leqslant W < 55$	12
$55 \leqslant W < 60$	10
$60 \leqslant W < 65$	6
$65 \leqslant W < 70$	2

a) Calculate an estimate of the mean weight of the students.

....

....

....

.... ④

b) Which class interval contains the median weight?

....

.... ②

TOTAL 6

---

18 Gilly carried out a survey on the number of hours spent on homework in a week. She asked 100 students in her school.

Her results are shown in the table below:

Hours (x)	Frequency		Cumulative frequency
$0 < x \leqslant 2$	3	$x \leqslant 2$	
$2 < x \leqslant 4$	7	$x \leqslant 4$	
$4 < x \leqslant 6$	13	$x \leqslant 6$	
$6 < x \leqslant 8$	25	$x \leqslant 8$	
$8 < x \leqslant 10$	32	$x \leqslant 10$	
$10 < x \leqslant 12$	12	$x \leqslant 12$	
$12 < x \leqslant 14$	6	$x \leqslant 14$	
$14 < x \leqslant 16$	2	$x \leqslant 16$	

a) Complete the cumulative frequency table for the 100 students. ①

b) Draw the cumulative frequency diagram on the grid below.

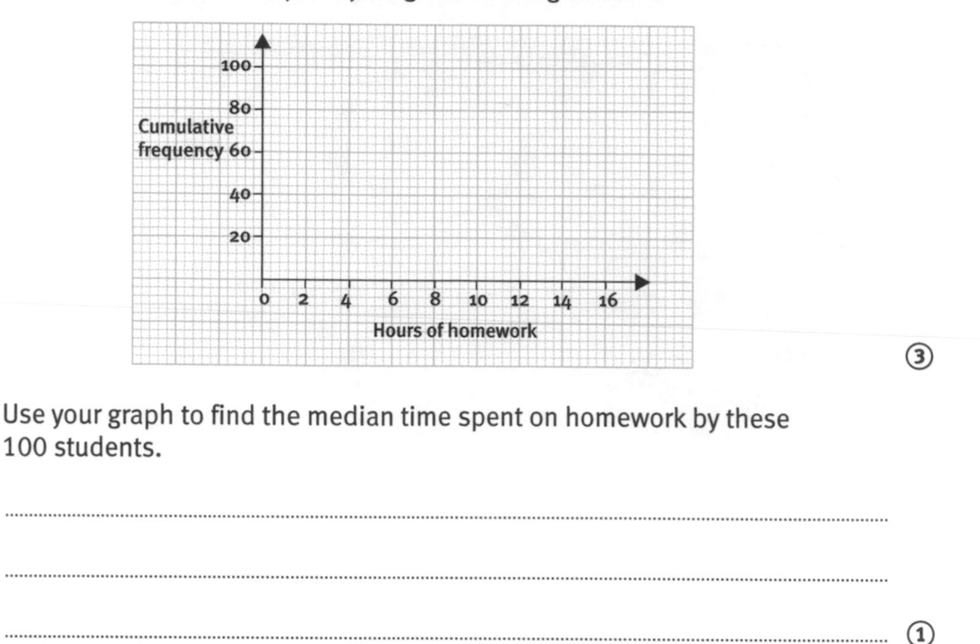

③

c) Use your graph to find the median time spent on homework by these 100 students.

....

....

.... ①

d) Use your graph to work out an estimate for the Interquartile range.

....

.... ②

e) Use your graph to estimate how many of these 100 students did more than 9 hours homework in the week.

....

....

.... ②

TOTAL 9

19 The table shows the time in minutes for 95 people's journey times to work.

Time in minutes (t)	Frequency		Cumulative frequency
$0 \leq t < 10$	5	$t < 10$	
$10 \leq t < 20$	22	$t < 20$	
$20 \leq t < 30$	29	$t < 30$	
$30 \leq t < 40$	21	$t < 40$	
$40 \leq t < 50$	14	$t < 50$	
$50 \leq t < 60$	4	$t < 60$	

a) Complete the cumulative frequency column in the table above for the 95 people's journeys. ①

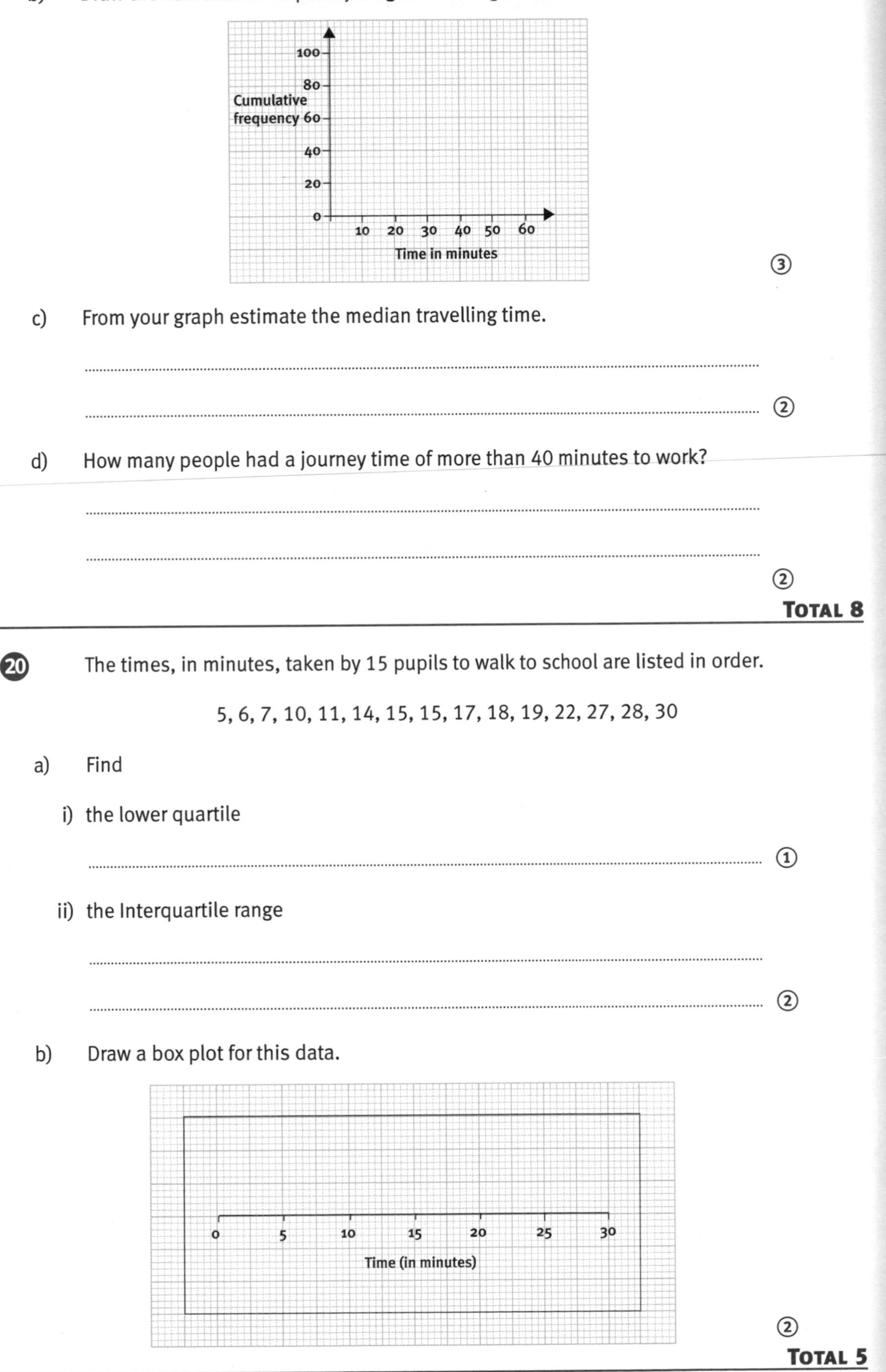

b) Draw the cumulative frequency diagram in the grid below.

③

c) From your graph estimate the median travelling time.

....

.... ②

d) How many people had a journey time of more than 40 minutes to work?

....

....

②

TOTAL 8

---

20 The times, in minutes, taken by 15 pupils to walk to school are listed in order.

5, 6, 7, 10, 11, 14, 15, 15, 17, 18, 19, 22, 27, 28, 30

a) Find

i) the lower quartile

.... ①

ii) the Interquartile range

....

.... ②

b) Draw a box plot for this data.

②

TOTAL 5

ANSWERS ON PAGE 78 ANSWERS ON PAGE 78 ANSWERS ON PAGE 78 ANSWERS ON PAGE 78

21 The data below shows the length, in centimetres, of 20 seedlings.

3.2	2.6	1.9	4.7	4.1	3.1	4.3	2.9	2.1	4.9
4.8	1.3	5.2	4.6	3.9	3.8	3.0	1.7	3.2	2.3

a) Make a stem and leaf table, 3|2 represents 3.2 cm.

③

b) Find the median length of the seedlings.

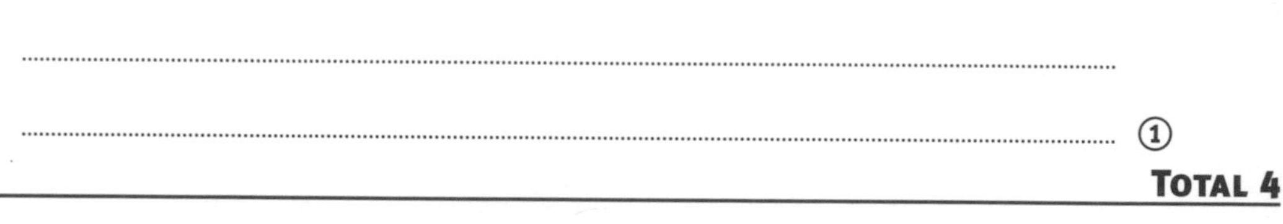

①

TOTAL 4

22 Rupinder is carrying out a survey of Year 11 pupils' weights in kilograms. The box plots show her data for boys and girls.

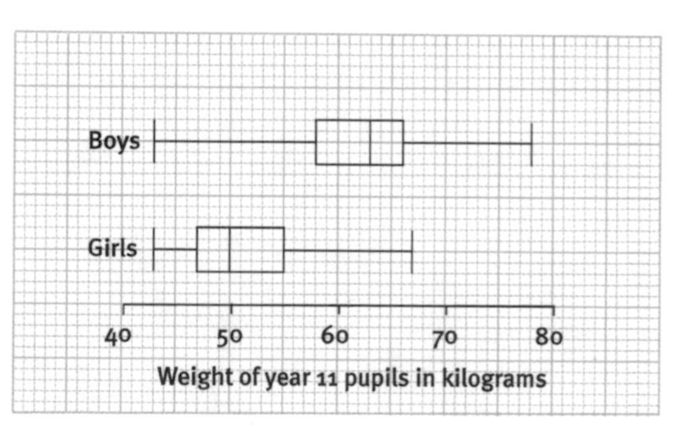

Make two comments about the data.

②

TOTAL 2

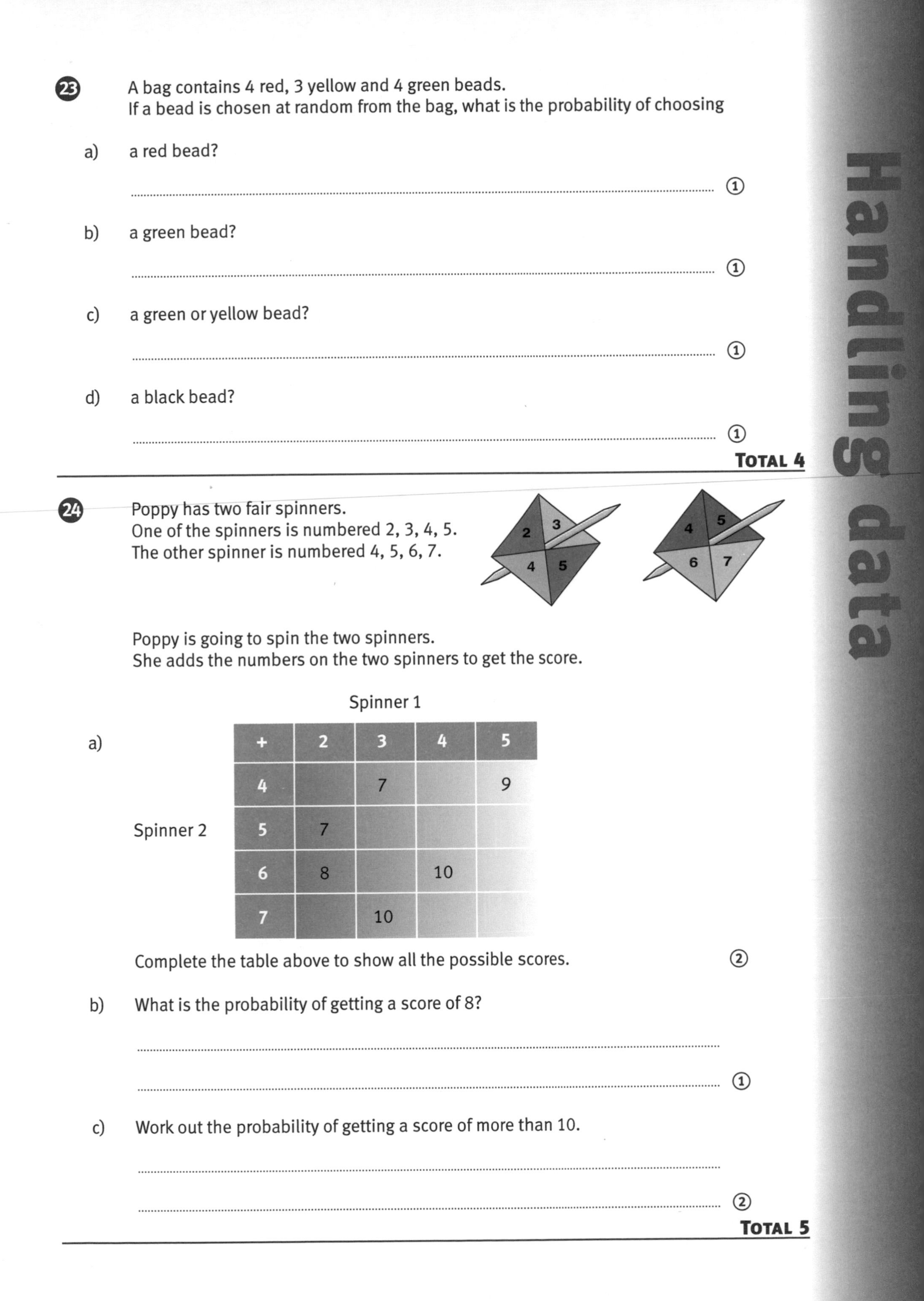

23 A bag contains 4 red, 3 yellow and 4 green beads.
If a bead is chosen at random from the bag, what is the probability of choosing

a) a red bead?

.... (1)

b) a green bead?

.... (1)

c) a green or yellow bead?

.... (1)

d) a black bead?

.... (1)

TOTAL 4

---

24 Poppy has two fair spinners.
One of the spinners is numbered 2, 3, 4, 5.
The other spinner is numbered 4, 5, 6, 7.

Poppy is going to spin the two spinners.
She adds the numbers on the two spinners to get the score.

a)

Spinner 1

Spinner 2 / +	2	3	4	5
4		7		9
5	7			
6	8		10	
7		10		

Complete the table above to show all the possible scores. (2)

b) What is the probability of getting a score of 8?

....

.... (1)

c) Work out the probability of getting a score of more than 10.

....

.... (2)

TOTAL 5

ANSWERS ON PAGE 78 ANSWERS ON PAGE 78 ANSWERS ON PAGE 78 ANSWERS ON PAGE 78

Here is a five-sided spinner.

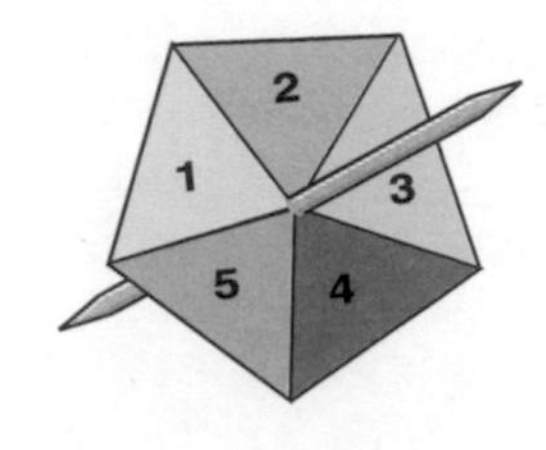

The sides are labelled 1, 2, 3, 4, 5.
The spinner is biased.

The probability that the spinner will land on each of the numbers 1 to 4 is given in the table.

Number	1	2	3	4	5
Probability	0.2	0.1	0.3	0.25	

Mario spins the spinner once.

a) i) Work out the probability that the spinner lands on the number 5.

....

.... ①

ii) Which number is the spinner least likely to land on?

....

.... ①

iii) Work out the probability that the spinner will land on a prime number.

....

....

.... ②

Jack spins the spinner 200 times.

b) Work out an estimate for the number of times the spinner will land on the number 3.

....

.... ②

c) i) Jessica spins the spinner twice. Work out the probability that the spinner lands on a 4 both times.

....

.... ②

ii) Work out the probability that Jessica will get a 1 and a 5.

....

.... ②

TOTAL 10

ANSWERS ON PAGE 78 ANSWERS ON PAGE 78 ANSWERS ON PAGE 78 ANSWERS ON PAGE 78

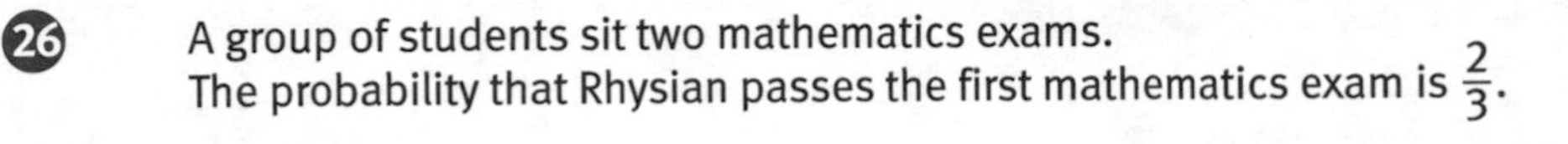

26 A group of students sit two mathematics exams.
The probability that Rhysian passes the first mathematics exam is $\frac{2}{3}$.

The probability that Rhysian passes the second mathematics exam is $\frac{4}{5}$.

a) Complete the tree diagram to show the outcomes when Rhysian sits both mathematics exams. The tree diagram has been started below:

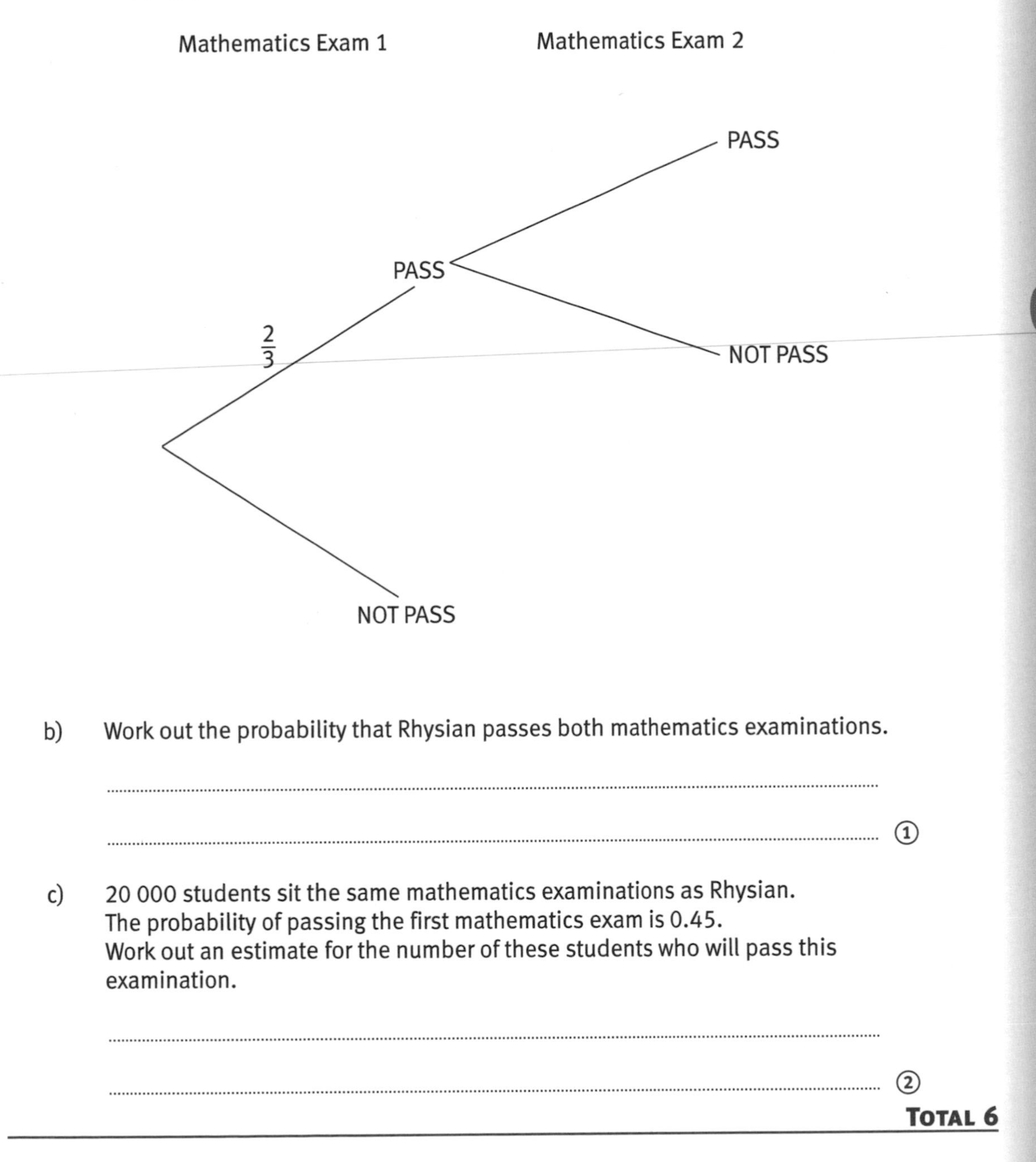

b) Work out the probability that Rhysian passes both mathematics examinations.

....

.... ①

c) 20 000 students sit the same mathematics examinations as Rhysian.
The probability of passing the first mathematics exam is 0.45.
Work out an estimate for the number of these students who will pass this examination.

....

.... ②

TOTAL 6

QUESTION BANK ANSWERS

1

Types of films	Tally	Frequency

❸

EXAMINER'S TIP

You need to have a heading called 'types of films', rather than just a list of names of films, to get the mark.

2 Reece's question is better than Charlotte's because it has a checklist that people can tick. Charlotte's question is too vague. ❸

EXAMINER'S TIP

When using boxes, check that the groups do not overlap.

3 a)

	Fiction	Non fiction	No books	Total
Boys	18	18	4	40
Girls	25	5	10	40
Total	43	23	14	80

❷

b) 10 girls ❶

4

	French	German	Spanish	Total
Year 10	38	42	35	115
Year 11	63	51	21	135
Total	101	93	56	250

❷

b) 93 students ❶

EXAMINER'S TIP

In two way tables, the totals for the rows should be the same as the totals for the columns.

5

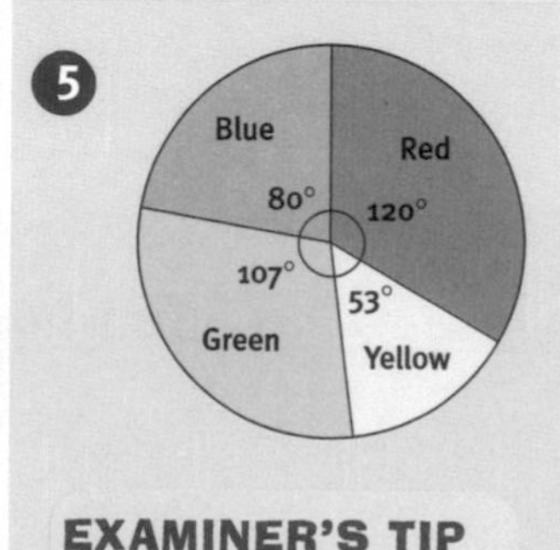

❹

EXAMINER'S TIP

Check that the angles are measured accurately to within 2 degrees.

6 a) $\frac{135°}{360°} \times 24 = 9$ hours ❷

b) 5 hours ❷

7 a)

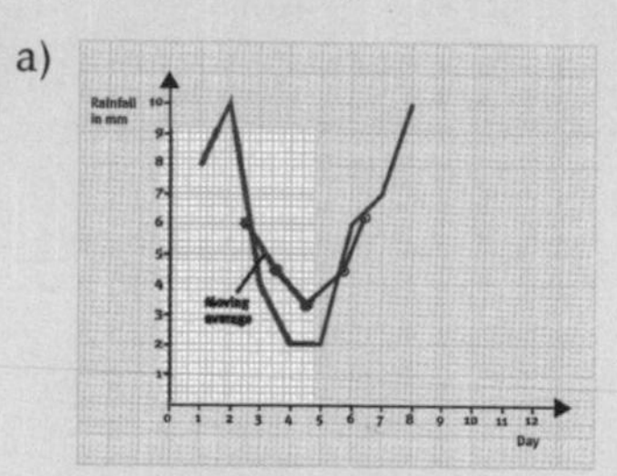

❹

b)

Day	mm of Rainfall	Moving Average
1	8	
2	10	
3	4	6
4	2	4.5
5	2	3.5
6	6	4.25
7	7	6.25
8	10	

❸

c) see graph ❸

d)
- General trend drops and then begins to rise.
- Amount of rainfall drops suddenly during days 3 to 5. ❷

EXAMINER'S TIP

Remember to plot the moving averages at the midpoint of each time period.

8

Key 1|2 means 12 mins

0	
1	2 7 7 8
2	3 3 5 6 7 7
3	3 4 6 6 8
4	1 2 9
5	2 7

❹

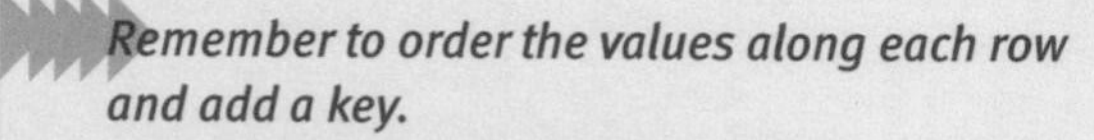

EXAMINER'S TIP

Remember to order the values along each row and add a key.

9

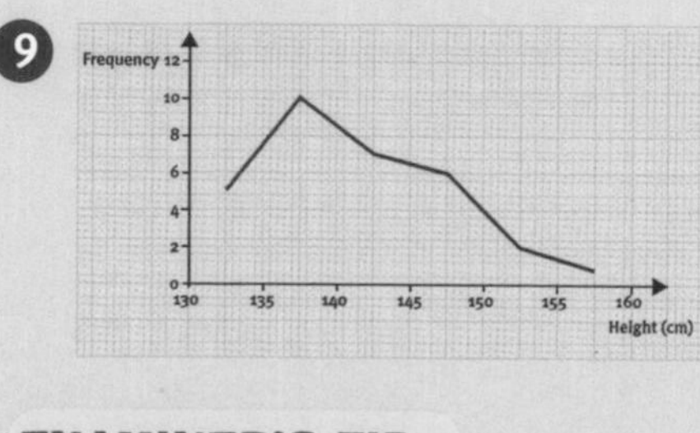

❷

EXAMINER'S TIP

Remember to plot the midpoints of the class intervals.

FOR MORE INFORMATION ON THIS TOPIC ... SEE GCSE SUCCESS GUIDE ... PAGES 74–89

10 a)

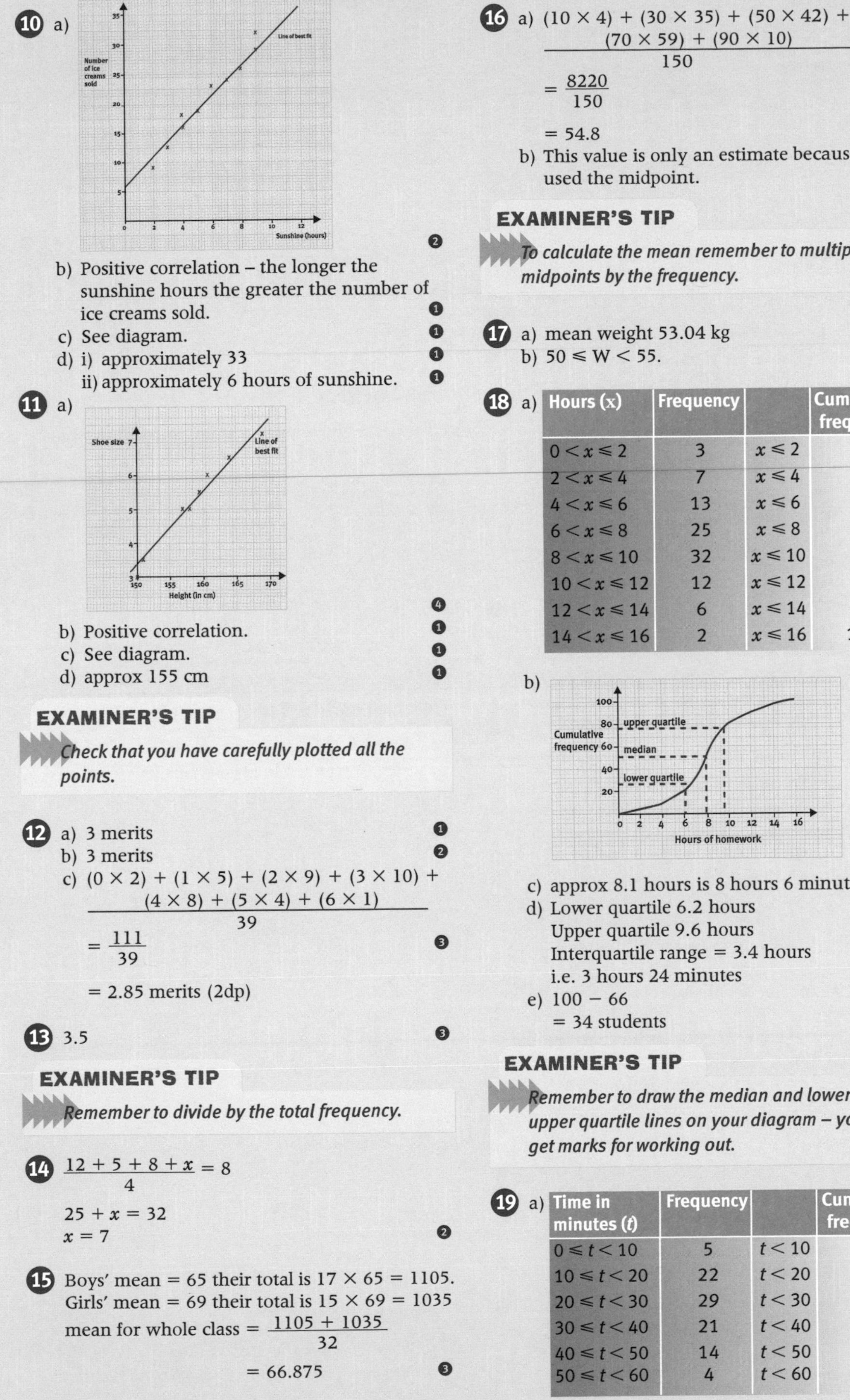

2

b) Positive correlation – the longer the sunshine hours the greater the number of ice creams sold. 1

c) See diagram. 1

d) i) approximately 33 1

ii) approximately 6 hours of sunshine. 1

11 a)

Shoe size / Line of best fit / Height (in cm)

4

b) Positive correlation. 1

c) See diagram. 1

d) approx 155 cm 1

EXAMINER'S TIP

Check that you have carefully plotted all the points.

12 a) 3 merits 1

b) 3 merits 2

c) $\frac{(0 \times 2) + (1 \times 5) + (2 \times 9) + (3 \times 10) + (4 \times 8) + (5 \times 4) + (6 \times 1)}{39}$

$= \frac{111}{39}$ 3

$= 2.85$ merits (2dp)

13 3.5 3

EXAMINER'S TIP

Remember to divide by the total frequency.

14 $\frac{12 + 5 + 8 + x}{4} = 8$

$25 + x = 32$

$x = 7$ 2

15 Boys' mean = 65 their total is $17 \times 65 = 1105$.
Girls' mean = 69 their total is $15 \times 69 = 1035$

mean for whole class $= \frac{1105 + 1035}{32}$

$= 66.875$ 3

16 a) $\frac{(10 \times 4) + (30 \times 35) + (50 \times 42) + (70 \times 59) + (90 \times 10)}{150}$

$= \frac{8220}{150}$

$= 54.8$ 4

b) This value is only an estimate because we used the midpoint. 1

EXAMINER'S TIP

To calculate the mean remember to multiply the midpoints by the frequency.

17 a) mean weight 53.04 kg 4

b) $50 \leq W < 55$. 2

18 a)

Hours (x)	Frequency		Cumulative frequency
$0 < x \leq 2$	3	$x \leq 2$	3
$2 < x \leq 4$	7	$x \leq 4$	10
$4 < x \leq 6$	13	$x \leq 6$	23
$6 < x \leq 8$	25	$x \leq 8$	48
$8 < x \leq 10$	32	$x \leq 10$	80
$10 < x \leq 12$	12	$x \leq 12$	92
$12 < x \leq 14$	6	$x \leq 14$	98
$14 < x \leq 16$	2	$x \leq 16$	100

1

b)

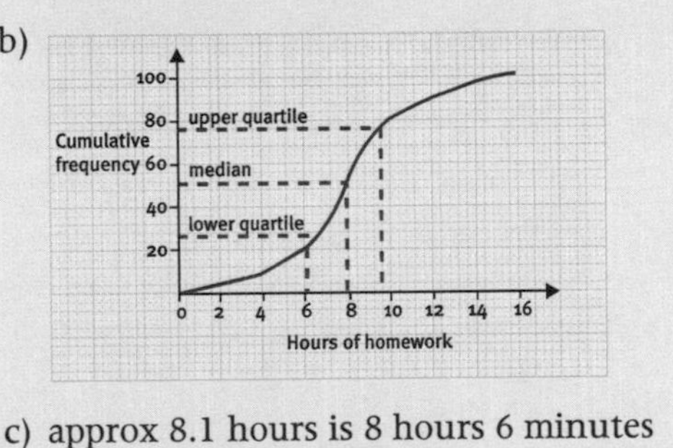

3

c) approx 8.1 hours is 8 hours 6 minutes 1

d) Lower quartile 6.2 hours
Upper quartile 9.6 hours
Interquartile range = 3.4 hours
i.e. 3 hours 24 minutes 2

e) $100 - 66$
$= 34$ students 2

EXAMINER'S TIP

Remember to draw the median and lower and upper quartile lines on your diagram – you may get marks for working out.

19 a)

Time in minutes (*t*)	Frequency		Cumulative frequency
$0 \leq t < 10$	5	$t < 10$	5
$10 \leq t < 20$	22	$t < 20$	27
$20 \leq t < 30$	29	$t < 30$	56
$30 \leq t < 40$	21	$t < 40$	77
$40 \leq t < 50$	14	$t < 50$	91
$50 \leq t < 60$	4	$t < 60$	95

1

FOR MORE INFORMATION ON THIS TOPIC ... SEE GCSE SUCCESS GUIDE ... PAGES 74–89

Handling data

b)

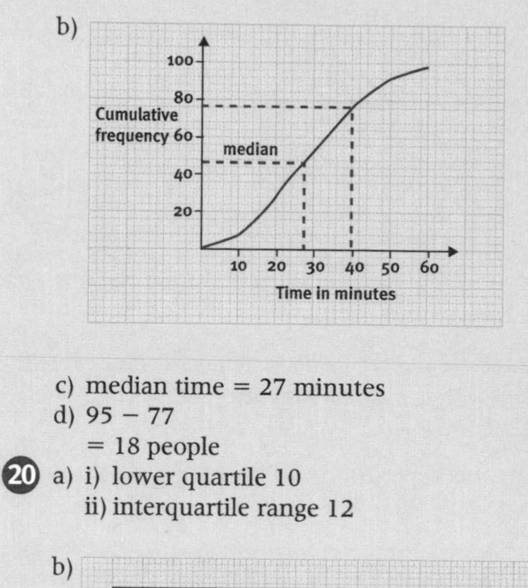

(3)

c) median time = 27 minutes (2)

d) 95 − 77
= 18 people (2)

20 a) i) lower quartile 10 (1)
ii) interquartile range 12 (2)

b)

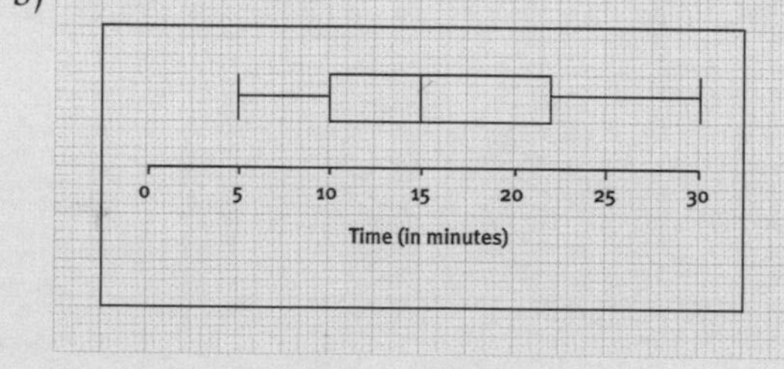

(2)

EXAMINER'S TIP

Remember that each section of the box plot represents 25% of the data.

21

1	3 7 9
2 1	3 6 9
3	0 1 2 2 8 9
4	1 3 6 7 8 9
5	2

Key 3|2 = 3.2 cm (3)

b) 3.2 cm (1)

22 e.g.

- The interquartile range for the boys is higher.
- The median for the girls is less than for the boys.
- The range for the boys is greater than for the girls.
- 50% of the girls weigh 50 kg or less compared with 50% of the boys who weigh 64 kg or less. (2)

23 a) $\frac{4}{11}$ (1)

b) $\frac{4}{11}$ (1)

c) $\frac{7}{11}$ (1)

d) 0 (1)

EXAMINER'S TIP

For part (d) it is best to write 0 and not 0/11. Remember to write probabilities as either a fraction, decimal or percentage.

24 a)

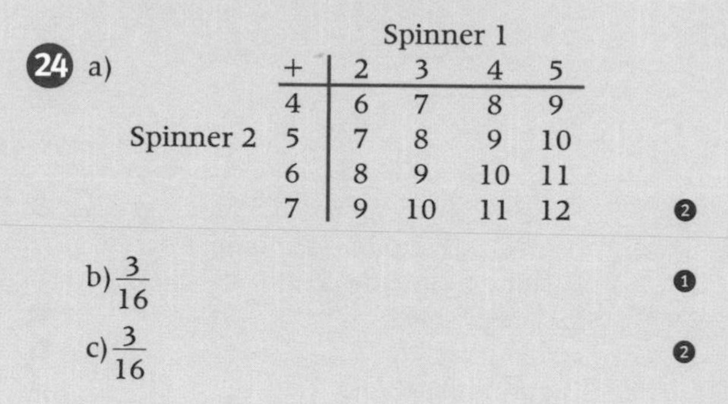

		Spinner 1			
	+	2	3	4	5
Spinner 2	4	6	7	8	9
	5	7	8	9	10
	6	8	9	10	11
	7	9	10	11	12

(2)

b) $\frac{3}{16}$ (1)

c) $\frac{3}{16}$ (2)

EXAMINER'S TIP

For part (c) remember it says more than 10.

25 a) i) 0.15 (1)
ii) 2 (1)
iii) 0.55 (2)

b) 0.3 × 200
= 60 (2)

c) i) 0.25 × 0.25 = 0.0625. (2)
ii) (0.2 × 0.15) + (0.15 × 0.2)
= 0.03 + 0.03
= 0.06 (2)

EXAMINER'S TIP

For part (c) (ii) remember to work out both alternatives and then add them together.

26 a)

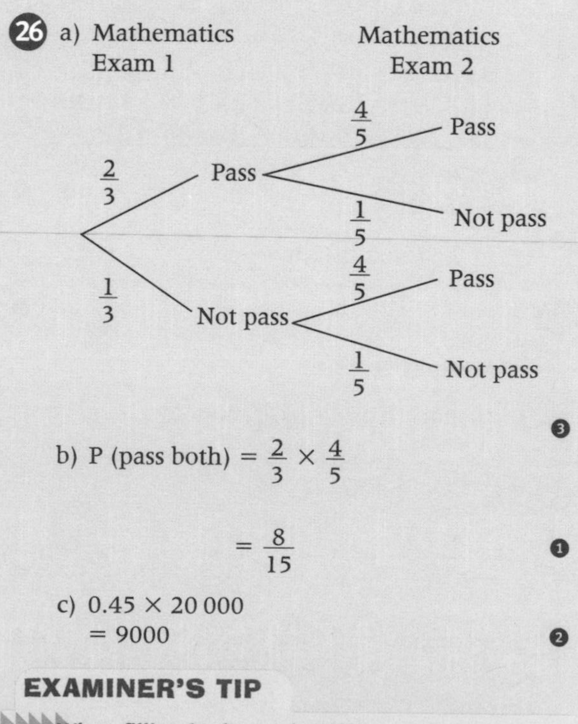

(3)

b) P (pass both) $= \frac{2}{3} \times \frac{4}{5}$

$= \frac{8}{15}$ (1)

c) 0.45 × 20 000
= 9000 (2)

EXAMINER'S TIP

When filling in the probabilities on the tree diagram, check that the branches add up to 1.

FOR MORE INFORMATION ON THIS TOPIC ... SEE GCSE SUCCESS GUIDE ... PAGES 74–89

Centre number
Candidate number
Surname and initials

Letts Examining Group
General Certificate of Secondary Education

Mathematics
Paper 1
Intermediate tier

Time: two hours

Instructions to candidates

Write your name, centre number and candidate number in the boxes at the top of this page.

Answer ALL questions in the spaces provided on the question paper.

Show all stages in any calculations and state the units. **You must not use a calculator in this paper**.

Include diagrams in your answers where this may be helpful.

Information for candidates

The number of marks available is given in brackets **[2]** at the end of each question or part question.

The marks allocated and the spaces provided for your answers are a good indication of the length of answer required.

For Examiner's use only	
1	
2	
3	
4	
5	
6	
7	
8	
9	
10	
11	
12	
13	
14	
15	
16	
17	
18	
Total	

Leave blank

1 Work out

(a) 3.27×3.8

....

....

....

.... **[3]**

(b) $16.25 \div 2.5$

....

....

....

.... **[3]**

(c) $2^3 \times 4^2$

....

.... **[3]**

(Total 9 marks)

Leave blank

2 Here is a rectangle:

5 cm

3 cm

The length of the rectangle is 5 cm.
The width of the rectangle is 3 cm.

(a) (i) Work out the area of the rectangle.

....

....cm^2

(ii) Work out the perimeter of the rectangle.

....

.... cm **[2]**

The rectangle is enlarged by a scale factor of 3.5.

(b) Write down the length and width of the enlarged rectangle.

Length .... cm

Width .... cm **[2]**

(Total 4 marks)

3 Some students took a test. The table gives information about their marks.

Mark	Frequency
3	4
4	2
5	3
6	1

Work out the mean mark.

....

....

....

.... **[2]**

(Total 2 marks)

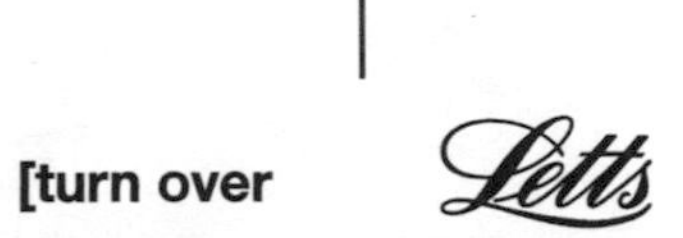

Leave blank

4 Construct an angle of 60° in the space provided.

[3]

(Total 3 marks)

5 The cost of 36 pencils is £6.48.

(a) Work out the cost of 24 pencils.

....

....

....

.... **[3]**

The probability that the pencil lead will break the first time it is used is 0.15.

(b) (i) Work out the probability that the lead of a pencil chosen at random will not break the first time it is used.

....

.... **[1]**

(ii) A box contains 300 pencils. Estimate the number of pencils whose lead will not break the first time it is used.

....

....

.... **[3]**

(Total 7 marks)

Leave blank

6 A cube has a surface area of 54 cm^2.

Work out the volume of the cube.

....

....

....

....

.... **[4]**

(Total 4 marks)

7 David shares £520 between his children Matthew and Emily in the ratio 5 : 3.

(a) Work out how much each child receives.

....

....

....

.... Matthew £ ....

Emily £ .... **[3]**

(b) What fraction of the £520 does Emily receive?

....

....

.... **[2]**

(Total 5 marks)

[turn over

8

Leave blank

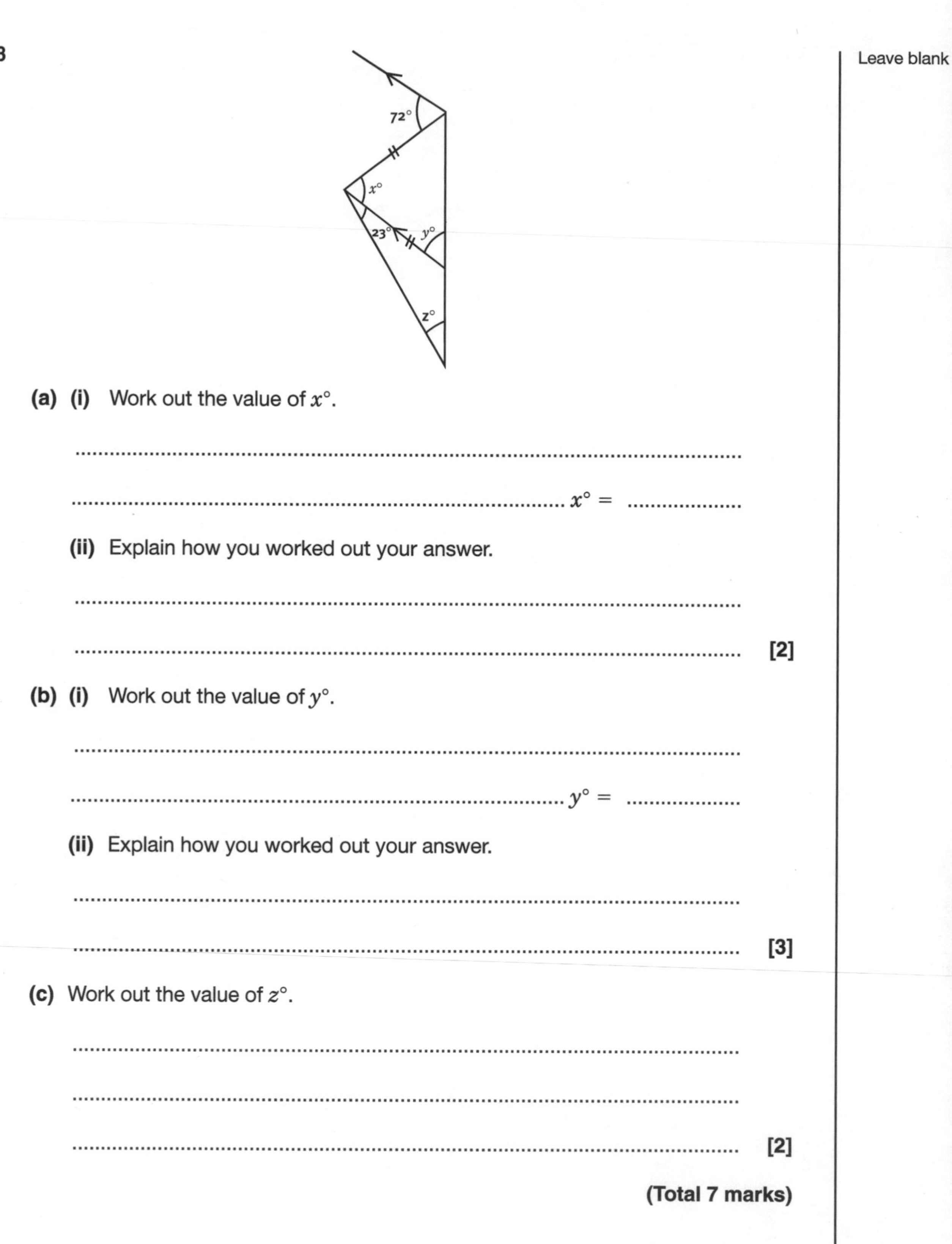

(a) **(i)** Work out the value of $x°$.

....

.... $x° =$ ....

(ii) Explain how you worked out your answer.

....

.... **[2]**

(b) **(i)** Work out the value of $y°$.

....

.... $y° =$ ....

(ii) Explain how you worked out your answer.

....

.... **[3]**

(c) Work out the value of $z°$.

....

....

.... **[2]**

(Total 7 marks)

9 Solve these equations.

Leave blank

(a) $5x = 20$

....

.... $x =$ .... **[1]**

(b) $2w - 3 = 9$

....

....

.... $w =$ .... **[2]**

(c) $5(2y - 1) = 25$

....

....

....

.... $y =$ .... **[2]**

(d) $8r - 5 = 3r + 20$

....

....

....

.... $r =$ .... **[3]**

(e) $\dfrac{5v - 3}{2v} = 4$

....

....

.... $v =$ .... **[3]**

(Total 11 marks)

10 Work out an estimate for the value of

$$\frac{62.5 \times 5.07 - 9.89 \times 3.06}{97.8^2}$$

Give your answer as a fraction in its simplest form.

....

....

....

.... **[3]**

(Total 3 marks)

11 The triangle R has been drawn on the grid.

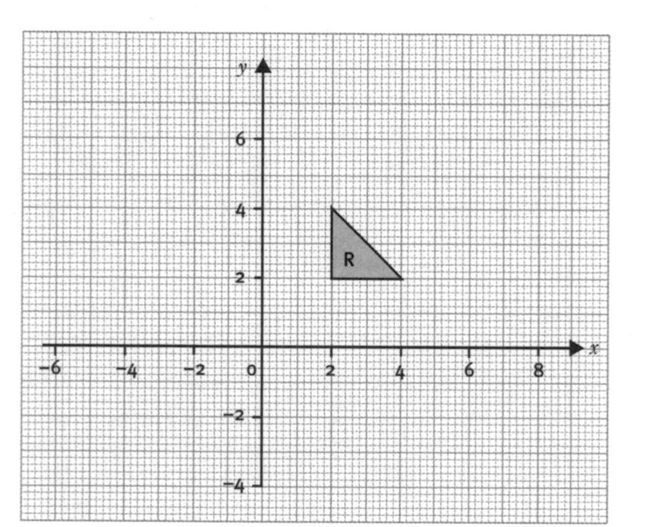

(a) Reflect the triangle R in the line $x = -1$
Label the image P.

[2]

(b) Rotate triangle R through 90° clockwise about (0, 0).
Label the image Q.

[2]

(c) Translate triangle R by the vector $\begin{pmatrix} -6 \\ -5 \end{pmatrix}$

Label the image S.

[2]

(Total 6 marks)

Leave blank

Leave blank

12 On the diagram draw the locus of the points outside the rectangle that are 2 cm away from the edges of this rectangle.

[3]

(Total 3 marks)

13 Two shops sell the same television set.

Robbie's
TV shop
£580 plus
VAT @ 17.5%

TV Mania
£760
with $^1/_5$ off.

(a) Work out the cost of the television sets in each of the two shops.

....

....

....

Robbie's TV shop ________

TV Mania ______________ [4]

(b)

ELECTRICAL
SUPERSTORE SALE
20% OFF
ALL TVs

In another shop the price of a different television set is £240 in the sale.

Work out the cost of the television set before the sale.

....

....

.... £ .... [4]

(Total 8 marks)

[turn over

Leave blank

14 n is an integer

(a) Write down the values of n which satisfy the inequality

$-4 < n \leqslant 2$

.... **[2]**

(b) Solve the inequality

$5p - 2 \geqslant 13$

....

.... **[2]**

(Total 4 marks)

15 The times, in minutes, taken by 11 people to wait to be served at a supermarket checkout are listed in order:

1, 1, 2, 3, 3, 3, 4, 5, 5, 6, 8

(a) Find

(i) the lower quartile

....

.... minutes

(ii) the interquartile range

....

.... **[2]**

(b) Draw a box plot for this data.

0 1 2 3 4 5 6 7 8 9 10

Time (in minutes)

[3]

(Total 5 marks)

Leave blank

16 $p = 6 \times 10^5$

$r = 3 \times 10^7$

(a) Find the value of $p \times r$.
Give your answer in standard form.

....

....

.... **[2]**

(b) Find the value of $\frac{p}{r}$.

Give your answer in standard form.

....

....

.... **[2]**

(Total 4 marks)

17 A, C and D are points on a circle, centre 0.

AB and BC are tangents to the circle.

Angle BCA is 74°.

Diagram not drawn to scale.

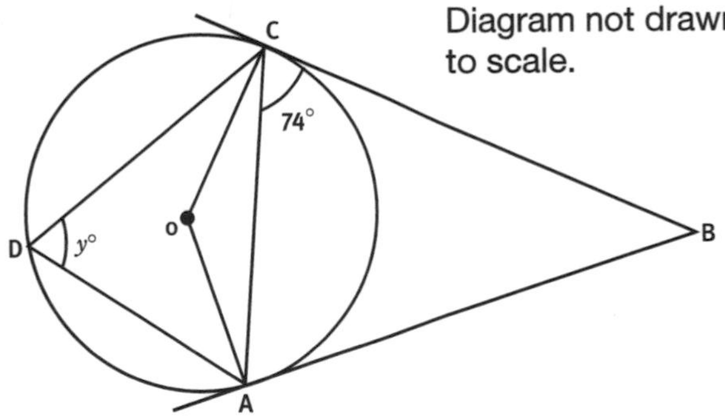

C

74°

D

$y°$

O

B

A

(a) Explain why angle OCA is 16°.

....

....

.... **[1]**

(b) Calculate the size of angle CDA.
Give reasons for your answer.

....

....

.... **[3]**

(Total 4 marks)

[turn over

Leave blank

18 **(a)** Simplify

$$y^5 \times y^3$$

.... **[1]**

(b) Expand and simplify

$$3(2x - 1) - 2(4x - 3)$$

....

.... **[2]**

(c) Expand and simplify

$$(a + b)^2$$

....

.... **[2]**

(d) **(i)** Factorise $6a - 3b$

....

....

(ii) Factorise completely

$$12p^2q - 9pq^2$$

....

.... **[3]**

(e) Find the value of

(i) 4^{-2}

....

(ii) 6^0

....

(iii) $16^{\frac{1}{2}}$

.... **[3]**

(Total 11 marks)

Centre number
Candidate number
Surname and initials

Letts Examining Group

General Certificate of Secondary Education

Mathematics
Paper 2
Intermediate tier

For Examiner's use only	
1	
2	
3	
4	
5	
6	
7	
8	
9	
10	
11	
12	
13	
14	
15	
16	
17	
18	
19	
20	
Total	

Time: two hours

Instructions to candidates

Write your name, centre number and candidate number in the boxes at the top of this page.

Answer ALL questions in the spaces provided on the question paper.

Show all stages in any calculations and state the units. You are expected to use a calculator in this paper.

Include diagrams in your answers where this may be helpful.

Information for candidates

The number of marks available is given in brackets **[2]** at the end of each question or part question.

The marks allocated and the spaces provided for your answers are a good indication of the length of answer required.

Leave blank

1 The data shows the birth weight, in kilograms, of 18 babies.

3.2	4.2	2.6	3.1	2.8	3.1	3.2	4.0	3.4
1.7	5.3	3.6	4.1	3.9	2.7	3.1	3.9	2.7

(a) Draw a stem and leaf diagram, where 4|2 represents 4.2.

[3]

(b) Find the median weight.

.... kg **[1]**

(Total 4 marks)

2 Here are the first four terms of an arithmetic sequence:

5, 8, 11, 14

Find an expression, in terms of n, for the nth term of the sequence.

....

....

.... **[2]**

(Total 2 marks)

3 Hannah asked 24 students to name their favourite flavour of crisps. Her results are shown in the table below:

Favourite flavour of crisps	Frequency
Cheese	8
Salt 'n' vinegar	7
Beef	4
Bacon	3
Ready salted	2

Draw an accurate pie chart to show this information. Use the circle below.

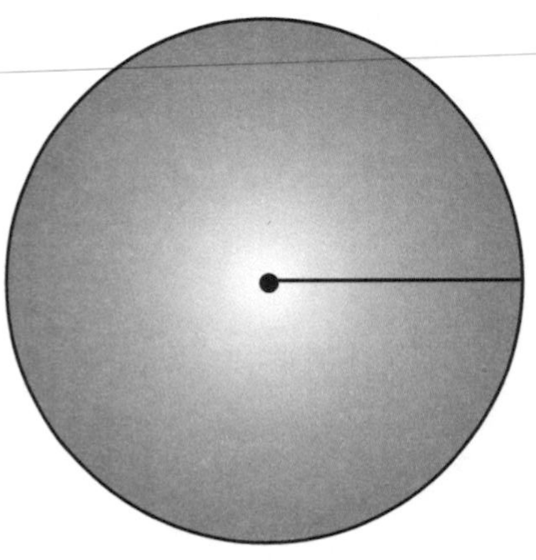

[4]

(Total 4 marks)

4 $p = 3q^2 + r$

(a) Find the value of p when $q = 2$ and $r = -3$.

....

....

....

.... **[2]**

(b) Rearrange the formula to make q the subject.

....

....

....

.... **[3]**

(Total 5 marks)

Leave blank

[turn over

5 Use the method of trial and improvement to solve the equation

$$t^3 + 2t = 39$$

Give your answer to 1 decimal place.
You must show all your working.

t = ... **[4]**

(Total 4 marks)

6 A circle has a radius of 5.5 cm.

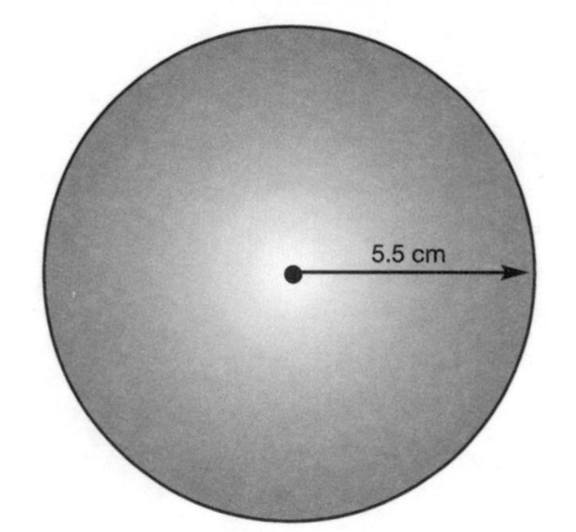

Diagram not drawn to scale.

(a) Work out the area of the circle.

Give your answer correct to 3 significant figures.

....

....

....

....

....

.... cm^2 **[2]**

Leave blank

(b)

Leave blank

Diagram not drawn to scale.

6 cm

8 cm

Work out the perimeter of the shape.
Give your answer correct to 3 significant figures.

....

....

....

.... cm **[3]**

(Total 5 marks)

7 Use your calculator to work out the value of

$$\frac{4.1^2 + 2.07\cos 30^\circ}{\sqrt{4.2 - 1.07}}$$

(a) Write down all the figures on your calculator display.

....

....

....

....

.... **[3]**

(b) Give your answer to part (a) to an appropriate degree of accuracy.

.... **[1]**

(Total 4 marks)

[turn over

Leave blank

8 **(a)** Simplify

$$3p - 2p + 4p - 6p$$

.... **[1]**

(b) Simplify

$$\frac{x^6}{x^2}$$

.... **[1]**

(c) Multiply out and simplify

$$(x - 2)(x - 4)$$

.... **[1]**

(d) Simplify

$$4x^4 \times 3x^2$$

.... **[1]**

(Total 4 marks)

9 **(a)** £8000 is invested for 3 years at 4% per annum compound interest. Work out the total interest earned over the three years.

....

....

....

....

.... **[4]**

(b) Colin buys a house for £142 000. Two years later he sells the house for £178 000. Work out his percentage profit.

....

....

....

....

.... **[3]**

(Total 7 marks)

Leave blank

10 The diagram shows a solid cylinder.

12 cm

9 cm

Diagram not drawn to scale.

(a) Calculate the volume of the cylinder. Give your answer correct to three significant figures.

....

....

....

....

....

....

....cm^3 **[3]**

(b) Write down the volume of the cylinder in m^3.
Leave your answer in standard form, correct to three significant figures.

....m^3 **[1]**

(Total 4 marks)

11 The grouped frequency table shows information about the number of hours spent travelling by each of 60 commuters in one week.

Number of hours spent travelling (t)	Frequency
$0 < t \leqslant 5$	0
$5 < t \leqslant 10$	14
$10 < t \leqslant 15$	21
$15 < t \leqslant 20$	15
$20 < t \leqslant 25$	7
$25 < t \leqslant 30$	3

(a) Write down the modal class interval.

....

.... **[1]**

(b) Work out an estimate for the mean number of hours spent travelling by the commuters that week.

[4]

(c) Complete the cumulative frequency table.

Number of hours spent travelling (t)	Cumulative Frequency
$0 < t \leqslant 5$	0
$0 < t \leqslant 10$	
$0 < t \leqslant 15$	
$0 < t \leqslant 20$	
$0 < t \leqslant 25$	
$0 < t \leqslant 30$	

[1]

(d) On the grid below, draw a cumulative frequency graph for your table.

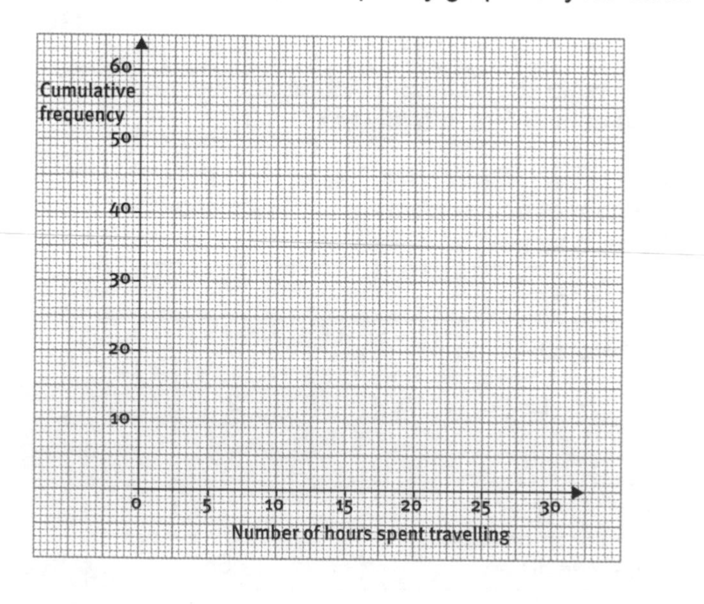

[2]

Leave blank

Leave blank

(e) Use your graph to find an estimate for the interquartile range of the number of hours spent travelling by the commuters that week.
Show your method clearly.

....

....

....hours **[3]**

(Total 11 marks)

12 **(a)** Express 54 and 36 as products of their prime factors.

....

....

....

54 ....

36 .... **[4]**

(b) Use your answer to (a) to work out the highest common factor of 54 and 36.

....

....

.... **[2]**

(Total 6 marks)

13 Solve the simultaneous equations

$$5v + 2w = 8$$
$$2v - w = 5$$

....

....

....

....

....

$v =$ ....

$w =$ .... **[4]**

(Total 4 marks)

[turn over

Leave blank

14 The table shows some expressions.
The letters p, q and r represent lengths.
π and 6 are numbers with no dimensions.
Two of the expressions could represent areas.
Tick the boxes under the two expressions which could represent areas.

$\frac{p^2q}{\pi}$	$r\sqrt{(p^2 - q^2)}$	$4(p^2 - r^2)$	$\frac{p^2qr}{4\pi}$	$p(q + r)^2$

[2]

(Total 2 marks)

15 (a) Complete the table of values for $y = 2x + \frac{1}{x}$.

x	0.1	0.2	0.4	0.6	0.8	1	2	3
y	10.2	5.4				3		6.3

[2]

(b) On the grid, draw the graph of $y = 2x + \frac{1}{x}$.

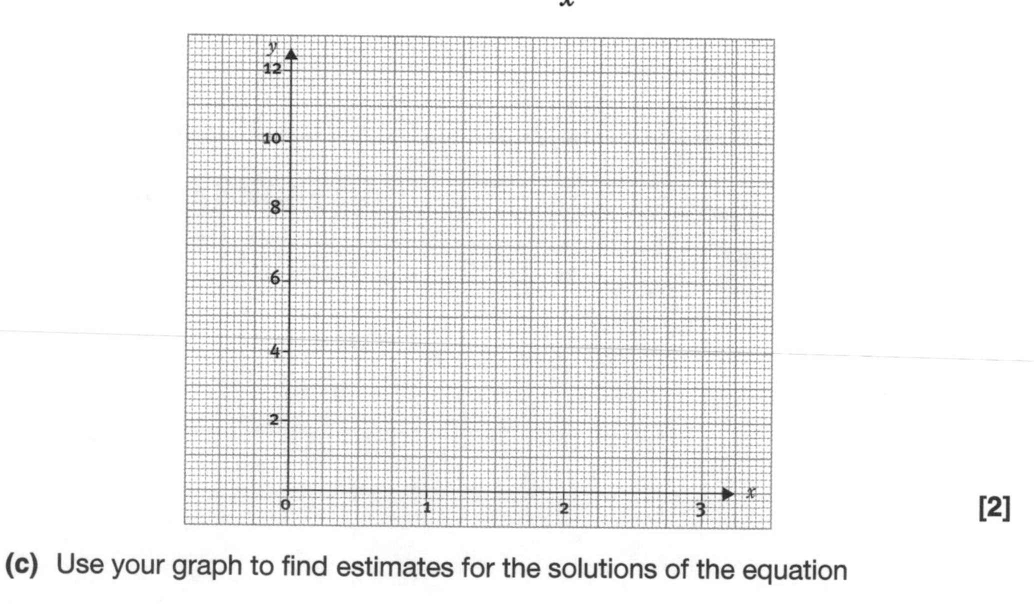

[2]

(c) Use your graph to find estimates for the solutions of the equation

$$2x + \frac{1}{x} = 5$$

$x =$ .... or $x =$ ....

[2]

(Total 6 marks)

Leave blank

16

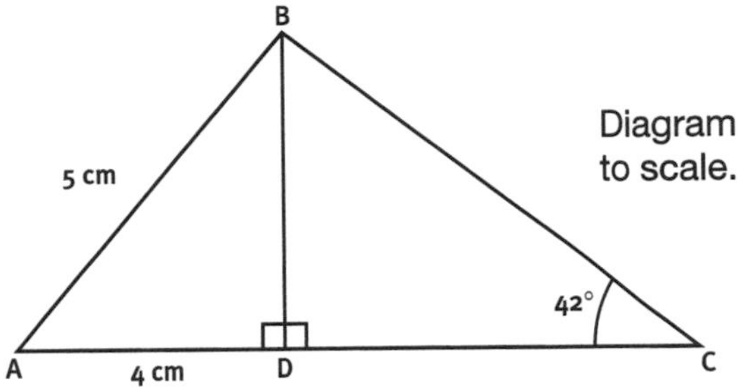

Diagram not drawn to scale.

The diagram shows two right-angled triangles: ABD and BCD.
The triangles are right angled at D.

AD = 4 cm
AB = 5 cm
Angle BCD = 42°

Calculate the length of BC.

....

....

....

....

....

.... **[5]**

(Total 5 marks)

[turn over

Letts

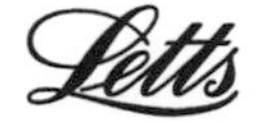

17

Leave blank

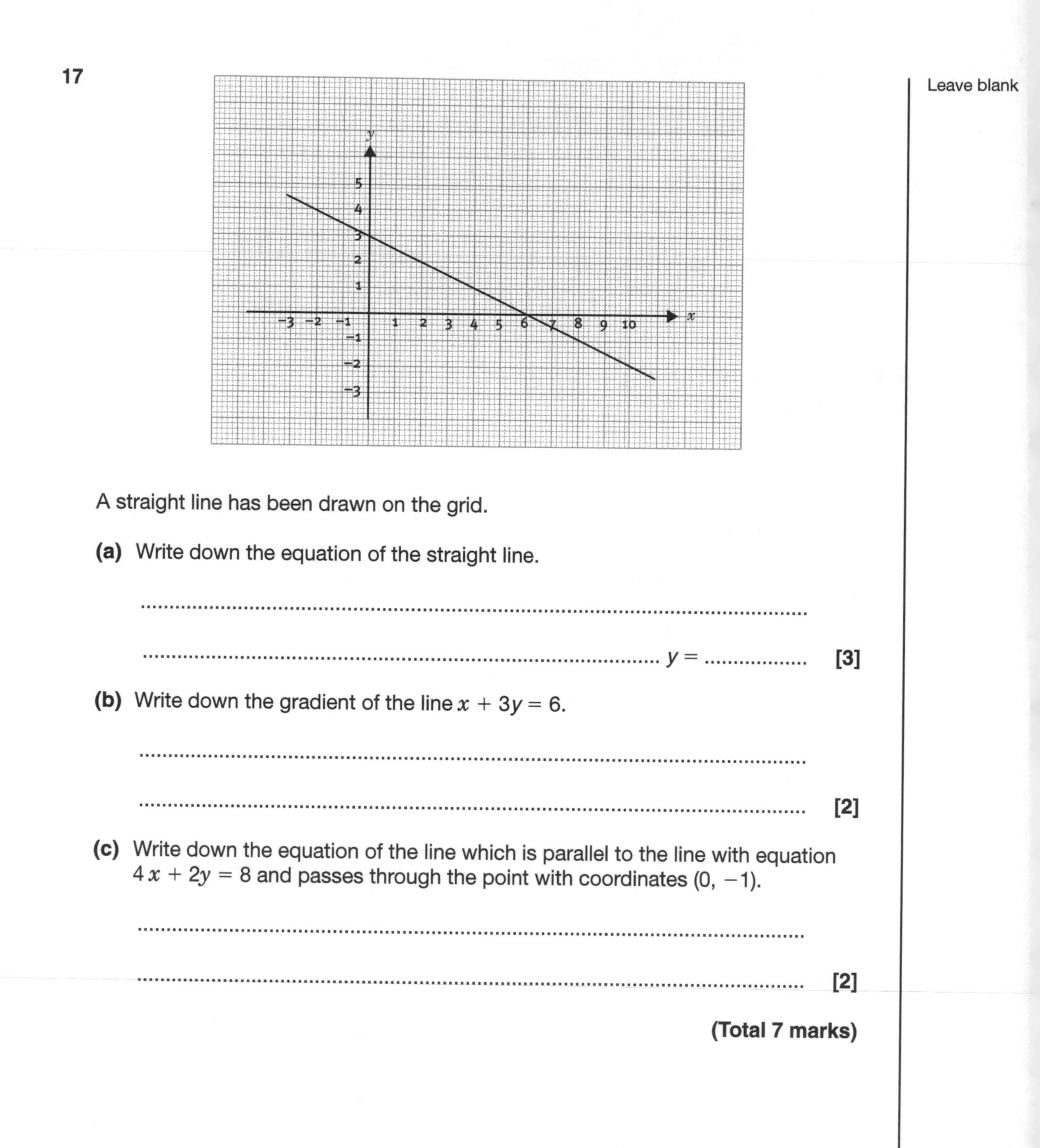

A straight line has been drawn on the grid.

(a) Write down the equation of the straight line.

....

.... $y =$ .... **[3]**

(b) Write down the gradient of the line $x + 3y = 6$.

....

.... **[2]**

(c) Write down the equation of the line which is parallel to the line with equation $4x + 2y = 8$ and passes through the point with coordinates $(0, -1)$.

....

.... **[2]**

(Total 7 marks)

18 Edward uses this formula to calculate the value of p.

$$p = \frac{a^2 b}{b - a}$$

Calculate the value of p when

$a = 4.5 \times 10^{-2}$ $\qquad$ $b = 8.6 \times 10^{7}$

Give your answer in standard form, correct to three significant figures.

....

....

.... **[3]**

(Total 3 marks)

19

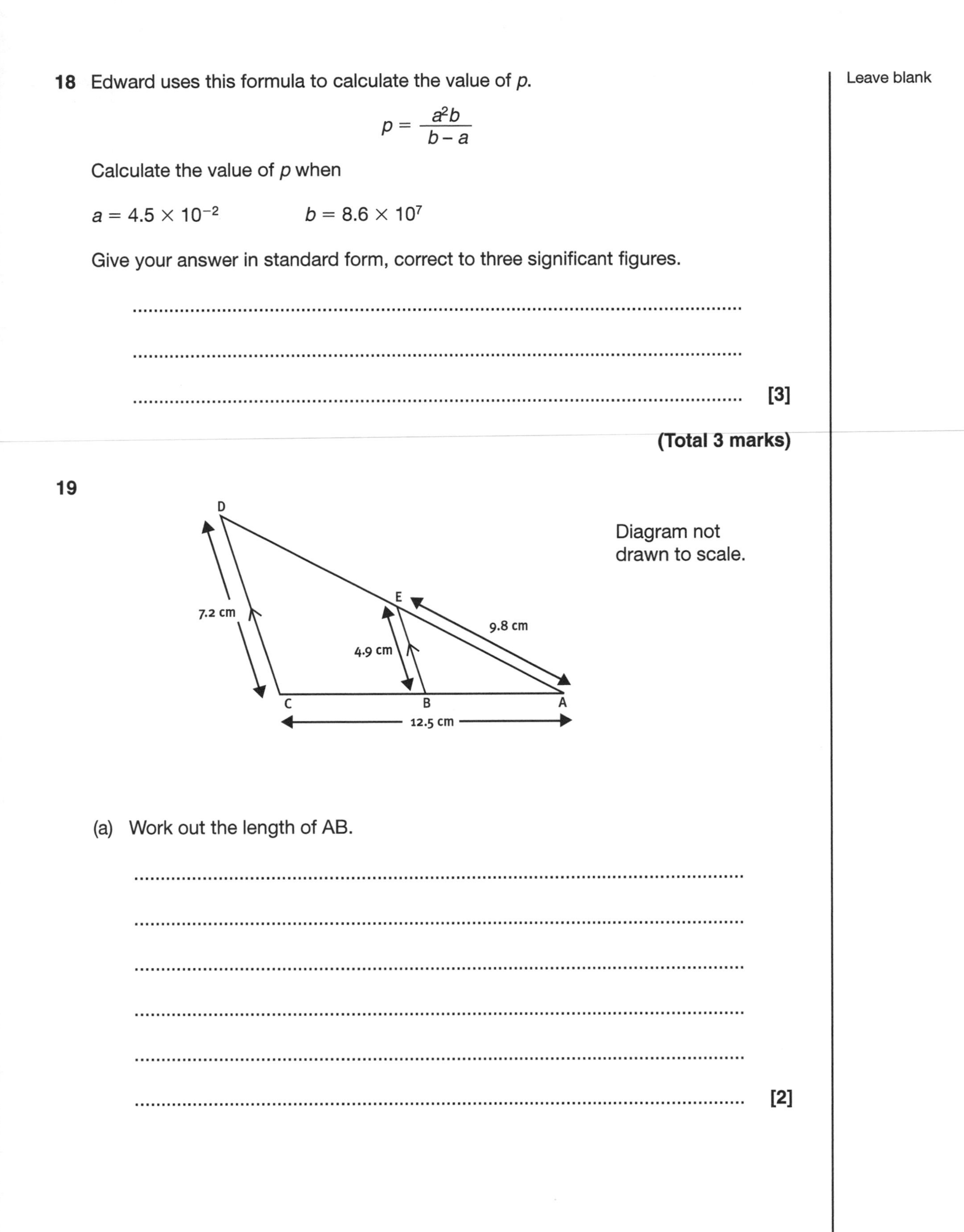

Diagram not drawn to scale.

(a) Work out the length of AB.

....

....

....

....

....

.... **[2]**

Leave blank

Leave blank

(b) Work out the length of DE.

....

....

.... cm **[2]**

(Total 4 marks)

20

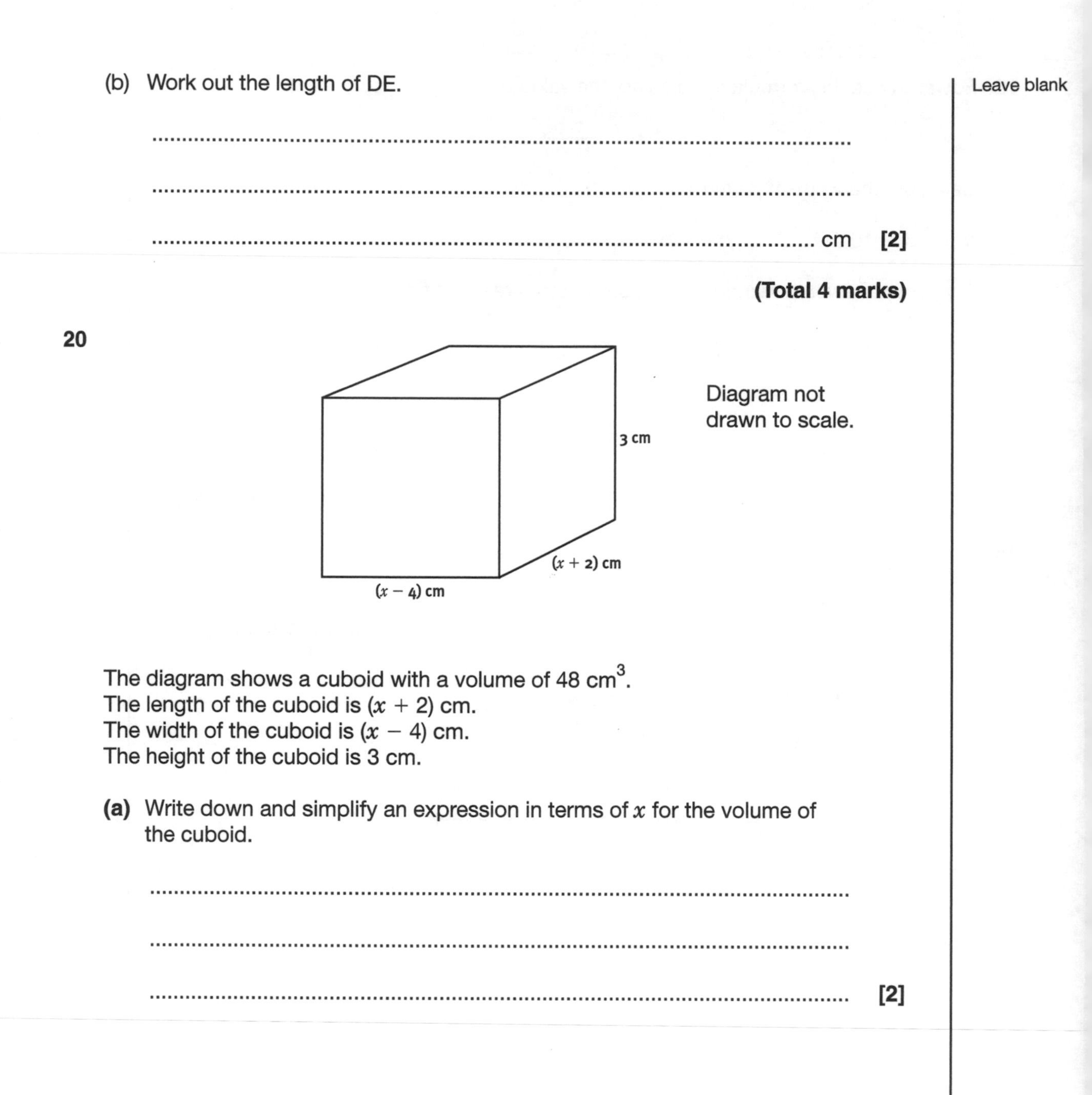

Diagram not drawn to scale.

The diagram shows a cuboid with a volume of 48 cm^3.
The length of the cuboid is $(x + 2)$ cm.
The width of the cuboid is $(x - 4)$ cm.
The height of the cuboid is 3 cm.

(a) Write down and simplify an expression in terms of x for the volume of the cuboid.

....

....

.... **[2]**

Given that the volume of the cuboid is 48 cm^3

Leave blank

(b) show that $x^2 - 2x - 24 = 0$

....

....

.... **[4]**

(c) Solve the quadratic equation and find the length and width of the cuboid.

....

....

....

Length = ...

Width = ... **[3]**

(Total 9 marks)

Answers to mock examination Paper 1

1 (a) 12.426 [3]
(b) 6.5 [3]
(c) 128 [3]

EXAMINER'S TIP

Remember 2^3 means $2 \times 2 \times 2$.

2 (a) (i) 15 cm^2
(ii) 16 cm [2]
(b) length 17.5 cm
width 10.5 cm [2]

3 4.1 [2]

EXAMINER'S TIP

Remember to divide by the sum of the frequencies.

4

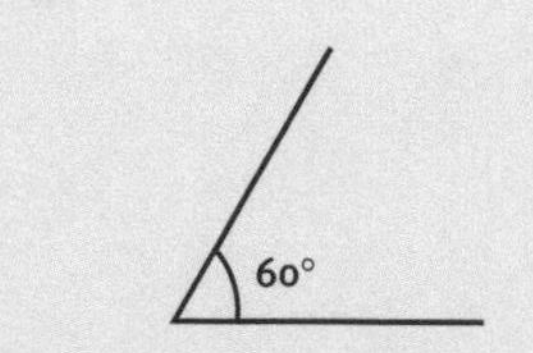

[3]

EXAMINER'S TIP

Constructing an angle of 60° is similar to constructing an equilateral triangle.

5 (a) £4.32 [3]
(b) (i) 0.85 [1]
(ii) $300 \times 0.85 = 255$ [3]

EXAMINER'S TIP

For part (b) (ii) it is easier to work out the number of pencils which will break and then subtract from 300.

6 27 cm^3 [4]

EXAMINER'S TIP

Work out the area of one face first then the length of the side.

7 (a) Matthew £325
Emily £195 [3]
(b) $\frac{3}{8}$ [2]

8 (a) (i) $x = 72°$
(ii) alternate angles with the 72° [2]
(b) (i) $y = 54°$
(ii) Base angles in an isosceles triangle are equal and angles in a triangle add up to 180°. [3]
(c) $z = 31°$ [2]

EXAMINER'S TIP

Always make explanations clear.

9 (a) $x = 4$ [1]
(b) $w = 6$ [2]
(c) $y = 3$ [2]
(d) $r = 5$ [3]
(e) $5v - 3 = 8v$
$-3 = 3v$
$v = -1$ [3]

EXAMINER'S TIP

For part (e) multiply both sides by 2v.

10 $\frac{(60 \times 5) - (10 \times 3)}{100^2}$

$= \frac{270}{10000}$

$= \frac{27}{1000}$ [3]

11 [6]

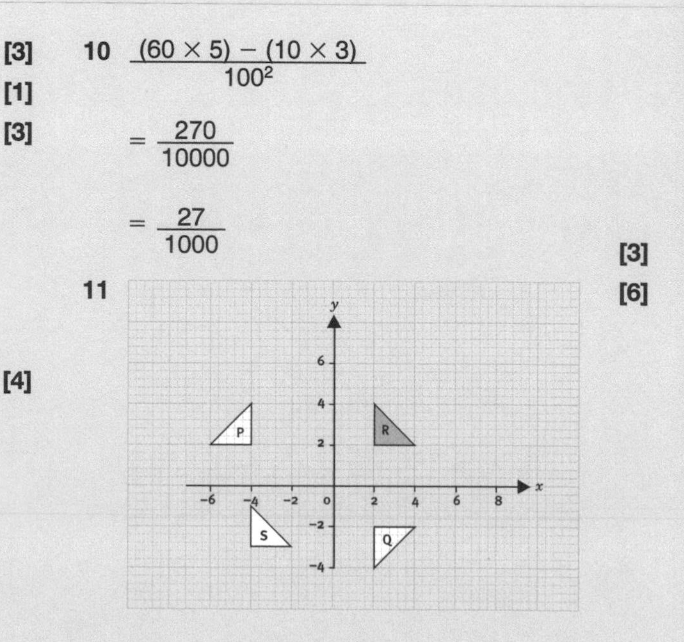

12

[3]

EXAMINER'S TIP

Remember to use a compass to form the arcs at the corners of the rectangle.

13 (a) Robbie's = £681.50
TV Mania = £608 **[4]**

(b) $\frac{240}{0.8} = \frac{2400}{8} = £300$ **[4]**

EXAMINER'S TIP

For part (b) remember that you are finding the original price so divide by the multiplier of 0.8.

14 (a) −3, −2, −1, 0, 1, 2 **[2]**

(b) $p \geqslant 3$ **[2]**

EXAMINER'S TIP

For part (a) remember not to include the −4.

15 (a) (i) lower quartile is 2 minutes

(ii) the interquartile range 3 minutes **[2]**

(b) **[3]**

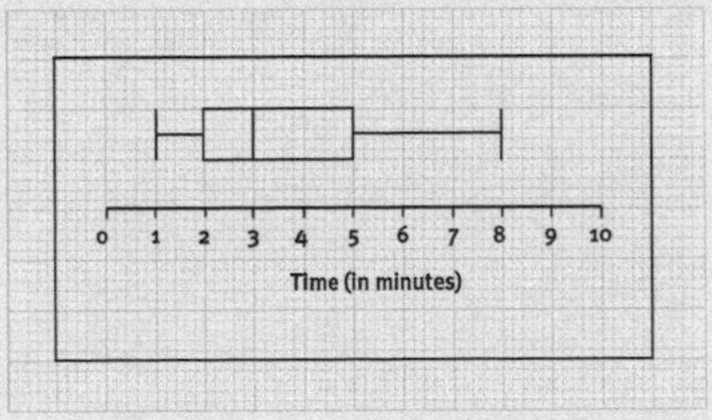

0 1 2 3 4 5 6 7 8 9 10
Time (in minutes)

16 (a) 1.8×10^{13} **[2]**

(b) 2×10^{-2} **[2]**

17 (a) OCA is 16° because the angle where a radius and tangent meet is 90°.
Hence 90° − 74° is 16°. **[1]**

(b) Angle OCA is 16°
Angle OAC is 16° } isosceles triangle
Hence angle AOC is 148°.
(Angles in a triangle add up to 180°.)
The angle at the circumference is half the angle at the centre.
Hence angle ADC is (148° ÷ 2).
ADC = 74° **[3]**

EXAMINER'S TIP

You need to give clear mathematical explanations to justify your answer.

18 (a) y^8 **[1]**

(b) $-2x + 3$ **[2]**

(c) $a^2 + 2ab + b^2$ **[2]**

(d) (i) $3(2a - b)$

(ii) $3pq(4p - 3q)$ **[3]**

(e) (i) $4^{-2} = \frac{1}{4^2} = \frac{1}{16}$

(ii) 1

(iii) $16^{\frac{1}{2}} = \sqrt{16} = +4$ or -4 **[3]**

EXAMINER'S TIP

Remember that $(a + b)^2$ *means* $(a + b)(a + b)$.

Answers to mock examination Paper 2

1 (a)

Stem	Leaf
1	7
2	6 7 7 8
3	1 1 1 2 2 4 6 9 9
4	0 1 2
5	3

Key 5/3 = 5.3 kg **[3]**

(b) 3.2 kg **[1]**

EXAMINER'S TIP

Remember to order the values in the stem and leaf diagram.

2 $3n + 2$ **[2]**

3

Cheese 120°
Salt 'n' vinegar 105°
Ready salted 30°
Bacon 45°
Beef 60°

[4]

EXAMINER'S TIP

Angles must be measured accurately to within plus or minus 2 degrees.

4 (a) $p = 9$ **[2]**

(b) $p = 3q^2 + r$

$q = \pm\sqrt{\frac{p \pm r}{3}}$ **[3]**

EXAMINER'S TIP

In part (a) remember to square the 2 first then multiply by 3.

5 $t = 3.2$ **[4]**

6 (a) 95.0 cm^2 **[2]**

(b) 32.6 cm **[3]**

EXAMINER'S TIP

You need to know the formulas for the area and circumference of a circle.

7 (a) 10.514852 **[3]**

(b) 10.5 **[1]**

8 (a) $-p$ **[1]**

(b) x^4 **[1]**

(c) $x^2 - 6x + 8$ **[1]**

(d) $12x^6$ **[1]**

9 (a) $8000 \times (1.04)^3$

$= 8998.91$

Interest = £998.91 **[4]**

(b) $\frac{36000}{142000} \times 100\% = 25.4\%$ **[3]**

EXAMINER'S TIP

When calculating the percentage profit remember to divide by the original value.

10 (a) Volume = 1020 cm^3 **[3]**

(b) Volume = 1.02×10^{-3} m^3 **[1]**

11 (a) $10 < t \leq 15$ **[1]**

(b) 14.5 hours **[4]**

(c)

Number of hours spent travelling (t)	Cumulative frequency
$0 < t \leq 5$	0
$0 < t \leq 10$	14
$0 < t \leq 15$	35
$0 < t \leq 20$	50
$0 < t \leq 25$	57
$0 < t \leq 30$	60

[1]

(d)

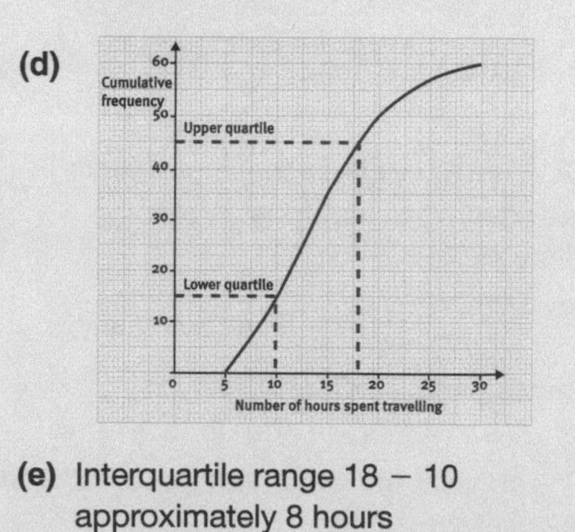

[2]

(e) Interquartile range 18 − 10 approximately 8 hours **[3]**

EXAMINER'S TIP

For part (e) it is useful to show the lower and upper quartile lines.

12 (a) $54 = 2 \times 3^3$
$36 = 2^2 \times 3^2$ **[4]**

(b) HCF = 18 **[2]**

13 $v = 2$
$w = -1$ **[4]**

14

$\frac{p^2q}{\pi}$	$r(\sqrt{p^2 - q^2})$	$4(p^2 - r^2)$	$\frac{p^2qr}{4\pi}$	$p(q + r)^2$
	✓	✓		

[2]

15 (a)

x	0.1	0.2	0.4	0.6	0.8	1	2	3
y	10.2	5.4	3.3	2.87	2.85	3	4.5	6.3

[2]

(b)

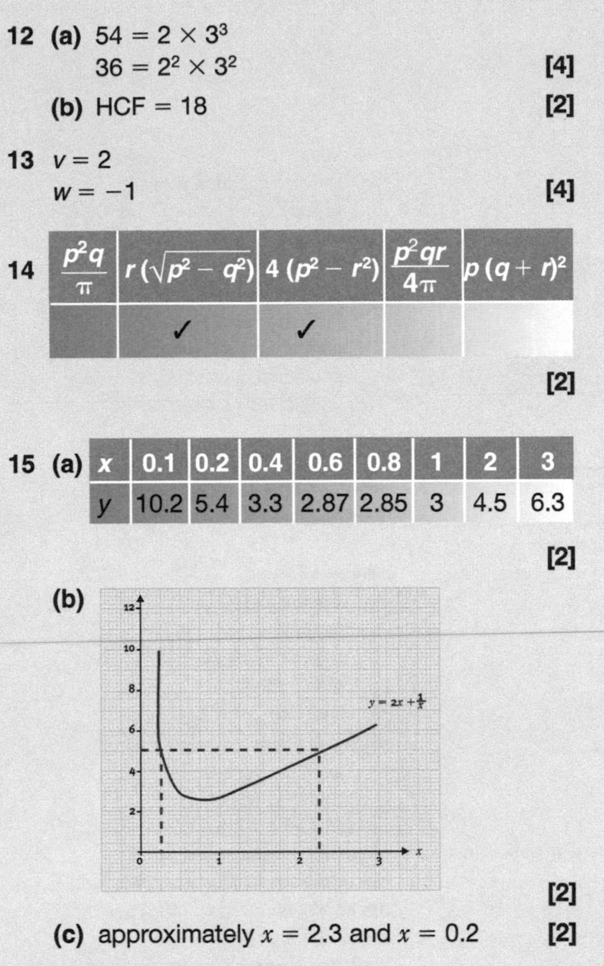

[2]

(c) approximately $x = 2.3$ and $x = 0.2$ **[2]**

EXAMINER'S TIP

To solve $2x + \frac{1}{x} = 5$, draw a line across from $y = 5$ and find the points of intersection – these are the required solutions.

16 4.48 cm (3sf) **[5]**

EXAMINER'S TIP

Firstly find BD by Pythagoras' theorem, then use the sine ratio to find BC.

17 (a) $y = -\frac{1}{2}x + 3$ **[3]**

(b) $-\frac{1}{3}$ **[2]**

(c) $4x + 2y = 8$
$2y = 8 - 4x$
$y = 4 - 2x$
equation of the line is $y = -2x - 1$ **[2]**

EXAMINER'S TIP

It is helpful to write the equation in the form y = mx + c *when answering questions like these.*

18 2.025×10^{-3} **[3]**

19 (a) AB = 8.51 cm **[2]**

(b) DE = 4.6 cm **[2]**

20 (a) $3(x - 4)(x + 2) =$ Volume **[2]**

(b) $3(x - 4)(x + 2) = 48$
$(x - 4)(x + 2) = 16$
$x^2 - 4x + 2x - 8 = 16$
$x^2 - 2x - 24 = 0$ **[4]**

(c) $x^2 - 2x - 24 = 0$
$(x + 4)(x - 6) = 0$
$x = -4$
$x = 6$
Using $x = 6$ Length = 8 cm
Width = 2 cm **[3]**

EXAMINER'S TIP

Remember when asked to solve a quadratic equation, make sure it is equal to zero, and then factorise.

HOW TO ASSESS YOUR GRADE

The grid below shows the grades that you might have expected to achieve with different scores on these papers. The marks are combined from paper 1 and paper 2 and are out of 200. No account has been made of the coursework marks. It is an indication only and does not imply that this is the grade that you will receive in the real examination.

Grades F and below are not awarded on the Intermediate Tier

Grade	Marks
B	132 – 200
C	89 – 131
D	49 – 88
E	0 – 48

Index